COWBOYS
You Gotta Love 'em

julie-carter.com

978-0-9741627-6-8

PPC+Publications
in cooperation with
Savvy Publishing, Estancia, NM

Cover cowboy is Bill Angell of Lovington, New Mexico, cover photograph is by Lara Studdard. Lara is a ranch wife, western photographer and mother of four daughters. When it comes to working cattle, Lara is alongside husband Dee or taking pictures of the working lifestyle of cowboys and cowgirls of the West.

Photography in this book is by Julie Carter.

FOREWORD

Following up a first book with a second book is like having a second child. The wonderment and awe of the event is no longer an unknown. You know that the birth of the child (book) will be exciting and comes with a new face, new personality and more adventures. You also know the work that is involved so you tend to pace yourself a little better.

I am very proud to "parent" this second book, part of an eventual series of three documenting the contemporary cowboy with tales of his everyday life. These stories go beyond the glamour of cowboy life perceived by many. The cowboys you meet within these pages are the real deal and their lives are as unglamorous as you can possibly imagine.

I promise you, my readers, a laugh, a moment of reflection and recall along with the new friends you will make with the characters in the stories. Welcome to my world, and in it, you will find a world much like your own, maybe with different names, places and jobs.

Julie Carter

DEDICATION

Like these stories and my books, my life is a work in progress. I dedicate this book to those cowboys and cowgirls that came before me, that formed who I am and taught me what I know. My father, my grandfather and my stepfather were all men of the West and of the land. From them I learned the solid values of Western living that serve as a foundation for me today. May my words do them honor. And, for all that I am, the glory belongs to God.

TABLE OF CONTENTS

Cowboy Tales

Good Stories and Bad Knees   8
Green Horses and a Red Wagon   10
Cowboy Multitasking   13
Technology for the Cowboy   16
What are Friends For?   18
New Twister Initiation   20
They Don't All Stay at the Ranch   23
Cowboys and Pot-Bellied Pigs are a Bad Mix   25
And the Horse He Rode in On   28
Young Buttons   31
Casey and the Grulla   34
Foreman of the Rough Ranch   37
Cowboy Lore Falls to New Low   40
Reject by eHarmony Dot Com   43
The Wall Street Cowboy   46
Horses and Snakes Top the Story List   49
Cowboys and ATVS, an Unnatural Mix   52
The Jingle of the Spur   55

Dan the Team Roper

A Tale of Slats, Dan and Whitey   60
Kid in the Farmer Brown Hat   62
It's Not Easy Being Frugally Cheap   65
Heelers by Any Other Name   68
Straw Hats and New Ropes   71
Doing It Wrong to Get It Right   74
All It Takes is a Good Ropin' Rock   77
Donnie and the Bunkhouse   80
The Tater Tot Explosion   83
Open Season on Winter Warmth   85
The Blame Game   88
The Devil is in the Details   91
You Can't Make This Stuff Up   93

Cowboy Slapstick is Alive and Well   96
Know When to Hold'Em, Know When to Run   99
Aristocrats of the Range   102

Horse Trading
The Art of Horse Trading   106
Vocabulary of a Horse Trader   109
Slipping Them By at the Sale Barn   111
Horse I Shouldn't Have Bought   114
$27,000 or Best Offer   117
Matching the Sales Pitch to the Horse   121

Panhandle Cowboys
Punching Cows in the Panhandle   126
I See By Your Outfit   129
Cowboys – Packaged Pride in What They Do   132

Ranch Wife Woes
Ranch Wife 201 – The Hundred-Year Dialogue   136
The Pre-apology Plan   139
Did I forget to Mention That?   142
Putting the Definition on Normal   145
That's Woman's Work   148
Cause and Effect   151

Rodeos, Ropers & Riders
The Bucket List   154
Sometimes the Best Cowboys are Cowgirls   157
Sally is a Good Ole Girl   160
Life After Rodeo   163
Call of the Rodeo Road   166
Old Glory Waves Across the Land Today   169
Barrel Racers – The Arena Darlins   172

Holidays
Cowboy Catalog Shopping   176
Gathering Strays to Sam's Place   178

Thankful is the Cowboy Way   181
The Christ In Christmas   183
Tis the Season for Romance for the Cowboy's Wife   186
Sadie Hawkins Cowboy-Style   190

Rural Fun
Summer Survival at the Annual Family Reunion   194
Return of the Hillbilly Family Reunion   197
Pigs, Kids and How to Meet Girls   200
The County Fair Belles of the Barn   203
Epic Clothesline Nostalgia   206
'Tis the Season for Camo and Ammo   209

COWBOY TALES

GOOD STORIES AND BAD KNEES

Cowboys, while being rugged individualists, have at least three things in common: a competitive spirit, the love of a good story and bad knees.

The knees are often the result of that aggressive nature and the stories most assuredly are.

One can quickly spot the cowboys without bad knees. As a rule, they are still attending high school classes. Through the years, their path can be traced through a number of orthopedic offices, culminating in the bionic replacement of body parts in later years. That is where the stories come in.

The gimping cowboy will willingly impart the story about that sun fishing son-of-a-gun that finally got the best of him and busted up that knee. A few of the more honest ones will admit to the limp originating with a football injury, even if that sometimes meant he fell out of the stands while watching.

Others will confess to less cowboy-like activities such as water or snow skiing or even a friendly game of Budweiser-fueled volleyball at a family reunion.

Cowboys with a horse- or cow-related limp will scoff at those embarrassing injuries and say with disdain, "It serves them right." One particular cowboy I know has cured the limp he acquired as a young cowboy and subsequently nursed all through his adult years as a cattleman. Then he became a cowboy again and took to team roping full time. The only benefit he can get from his limp now is bragging rights to the story.

He was day-working his way through college gathering cattle in the South Texas brasada. On the day of the legendary injury, he was assigned with a corrida of vaqueros to gather a bunch of snaky brush cattle.

They had spent the long, hot day in brush that consisted mostly of stickers and close-quartered oak and mesquite trees. Just as the cattle were finally gathered up and headed to the pens, one of the bulls decided he'd rather be where he had been rather than where he was.

The cowboy of this story and one other got the signal to go bring him back. The race was on. Both horses were fast and both men were hard riders. The brush was heavy and the bull thought his tail was on fire.

Our cowboy was in the lead to rope when the bull cut between two fair-size oaks. He calculated he had the one on the left cleared, shifted in the saddle to miss the one on the right.

The bull, the horse or the tree moved, it was never determined exactly which, and the cowboy hit the oak square on his knee. Of course, this happened as he was traveling at approximately the speed of light.

Nothing to say except, "Dang, that will hurt a feller."

After the knee healed up as good as it was ever going to, what hurt the most for a long time was that the other cowboy got to rope the bull. However, the now-gimpy cowboy had a wild tale he could tell for years. That almost made up for the bad knee.

Almost, except on cold mornings, long days in the saddle, long drives and, worse yet, when there was on one around to hear the tale.

GREEN HORSES AND A RED WAGON

Lots of things in a cowboy's life are not fair. Two items on the top of the inequitable list are having to feed in the mud when green grass should be filling that bill and fixing a water gap into a droughty pasture when the guy up the hill got all the rain.

In ranch country, feeding cattle is a standard operation that is driven by opposing weather conditions – the lack of rain or the abundance of snow.

Most outfits have the necessary equipment for feeding cattle in all kinds of weather.

The hands that operate it, and the cattle that depend on it, fall into a routine that continues throughout the season.

One year, this particular ranch had an unusually long rainy spell. Ranchers are known for their mantra of "never turn down a rain or a calf," but fighting the mud sometimes challenges their resolve on the "how much" rain issue.

About this time, two new hands had just hired on, green as the grass that had not yet grown and full of enthusiasm and energy. The deep mud had rendered the feed truck useless, but these two go-getters had seen pictures of ranchers feeding cattle with a wagon in some northern state.

Fortuitously, the ranch had a rubber-tired wagon, painted bright red, which was generally used for hayrides for the boss' kids. Perfect.

All they needed was a team of horses to pull it.

Their gaze turned to the large horse herd that supplied the saddle horses for the cattle operation. It seemed to them that would be a good place to find some wagon horses. Smart enough to know the saddle horses wouldn't know anything about being hitched to a wagon, they decided a couple of green colts would be "trainable" and wouldn't know any different. So they gathered up the 3-year-old colts, most of which had never been handled in any way.

They tied a back foot up on each colt and maneuvered them up to the wagon, where harnesses and collars were put in place while the colts stood trembling. A quick inspection of the wagon revealed a small problem. No brakes. The brighter of the two cowboys decided that a colt tied to the back of the wagon would work just fine for a brake.

When the driver (the honor given to the other cowboy) needed to stop, he could just wave his hat in the face of the colt at the back, making him set up and stop the wagon. Seemed like a fine idea, falling just short of brilliance.

The colts were wild-eyed and not happy about the wagon attached at their backsides. The lad on the ground untied their hind feet and when the driver slapped the reins on their backs, the colts stood momentarily frozen. Then, with a bolt, they left out like scalded hounds. The driver didn't need a whole lot of time to decide that the use of a brake might be in order.

He stood up on the wagon seat and waved his hat in the face of the colt tied to the back. As planned, the colt sat back hard. Not as planned, when the hard jerk hit the rope tied to the wagon, it pulled the entire back end out of the wagon and never even offered a moment of slowing to the run-off team. The driver, still attached to the fast-exiting horses, unwisely held tight to the reins while most of the hide on his face, and all the way south to his boots, was skidded off in the drag.

Finally, the colts managed to lose their dragger and, eventually, they ran out of steam. It still took most of the afternoon to get them caught and retrieve the harnesses. The green hands, not so green anymore, decided that God had created this mud and He would dry it out.

Wisdom is often born of pain or, in this case, rain. At least one cowboy is now a little wiser.

COWBOY MULTITASKING

Multitasking is a buzzword used to describe the art of doing several things simultaneously.

Cowboys are frequent multitaskers, especially when combining a little fun with their work.

Jess routinely worked at his brand of multitasking. This particular day, he was doctoring fresh yearlings and riding a green colt— always a recipe for a little excitement.

The big blue colt had kind eyes, a solid-built frame, long legs and a good heart. What he didn't have was any experience with cattle, a rope or a cowboy.

Jess was riding a brush pasture that bordered the Canadian River. The Panhandle has plenty of good grass under the mesquites and, in summer, the mesquite will yield a good crop of mesquite beans that pack the fat and sass on cattle and horses.

Once the yearlings were past the shipping fever stage, they'd be on the road to making a 300-pound gain. Money in the bank.

The colt, bought in early spring, would be a finished cow horse by first frost. That, too, would put a little more jingle in Jess' pockets. The plan was coming together.

Jess plow-reined the blue colt around mesquites and over downed cedars, teaching him how to place his feet. While the training was in progress, he watched for sick cattle lying out from the bunch.

The sick ones were roped easy and slow and then doctored. All the while, the colt was learning those skills under quiet hands. Jess got to feeling quite confident in his new horse about the same time he got tired of roping sick yearlings around the neck.

His thoughts wandered to the upcoming team roping in town and his need for practice roping horns.

Coincidentally, about that same time, a big, coming 3-year-old heifer showed up on his radar. She either had been missed on previous pasture gathers or had walked the river and taken up residence with the yearlings.

She had a nice spread of horns and would, to Jess' way of thinking, put the finishing touch on a good day for the colt.

Touching a spur to the colt, Jess built a loop, rode up on the heifer and roped her handily. He set the colt, stopped the heifer, dropped his rope down her right side, rode around her and let her front feet step over the loop and then pulled up the rope.

This was an easy way to put a half hitch on the back feet and lay her down. Jess rode the colt up to her, holding her back feet tight, reached down and took his rope off her horns. About that time, in rapid succession, the heifer came straight up, the colt went straight to the left, and Jess found himself standing on the ground with a heifer on the fight and a scared colt.

The heifer had never been roped and the colt had never seen a rope lying on the ground. The heifer's genetic make-up included plenty of old Mexican fighting-bull blood. She chased Jess around a mesquite while the colt eyed the rope like it was a monster. He didn't run, but he thought about it.

Eventually, the heifer got tired of the chase and walked off a little ways to catch her breath. Jess tried to ease up to the colt so as not to scare him into running. Headquarters was a long walk back.

About the time Jess got close to the colt, the heifer got a new spurt of mad and charged him around the mesquite a few more times. She repeated this for two hours. Finally, she walked off far enough for Jess to catch his colt, coil up the rope and get mounted.

Jess decided his next multitasking would include riding quietly back to headquarters and doing his roping practice in the arena.

It was obvious the lesson for the day was his.

TECHNOLOGY FOR THE COWBOY

When you see a cowboy leaning on the side of a pickup with an adult beverage in his hand, hat cocked back and a big toothy grin accenting the story he is telling, you just never, even remotely, consider there might be cutting-edge technology impacting his life, his job and his sport.

For the competitive roper, technology has become an integral part of his game.

For instance, a new rope has now come on the cowboy equipment scene that has some sort of space-age coating baked on it, so that it never loses its "slickum." Thereby, it is always fast in doing its job. This sci-fi layer replaces the wax coating common to good ropes.

There is also a rope available with a weighted tip in the loop. Evidently, someone, theoretically a roping consultant of reputed expertise, determines the perfect tip point. The rope is fashioned with a weight at that point. Bowling ball technology may have moved to the cowboy world.

The senior cowboys still hanging on in the new world of the techno-roper will regale you with stories of how ropes used to be made at home, not bought already manufactured.

A trip to the feed store allowed for the purchase of the right length of grass rope which was taken home, stretched for a millennium and left out to be weather-cured.

At just the right time, which often was determined by necessity, the rope was taken loose, a honda tied in it and a burner placed in the honda. A kitchen match struck on the backside of a denimed-leg was used to burn the burrs off the rope. With a bread wrapper or other such modern plastic technology, a slightly slick finish could be rubbed on the rope and it was good to go.

This piece of handcrafted equipment seemed to last longer. The amount of trouble it took to make it might have been a definite incentive for longevity. According to memory and legend, this masterpiece caught faster, bigger cattle more often with better accuracy. It could have also hung a few deserving people and in general was a valuable piece of equipment.

And as a bottom line, the initial grass rope to start this process cost only $4. The 100 hours of work put in the construction was valued at a quarter an hour. That brought the complete production cost to $29, which is, coincidentally, the same cost as today's no-count nylon and poly ropes. Progress is wonderful, isn't it?

Other technology gains have been as major as the invention of the horse trailer, pop-top cans eliminating the need for keeping track of the "church key" opener and the integration of sports physiology and sports psychology.

Video cameras allow for taping and analyzing runs to find places to shave off a hundredth of a second. It also gave the fence sitters a viable job instead of using the afternoon to empty the requisite resident cooler while telling the other ropers how it should be done.

This may account for some of the roping runs, which are over before you can say how much it would pay to win.

I'm a fairly technological person. I'm thinking I could easily upgrade to this more technological cowboy world.

In fact, I bet I could run the chute with that new electronic remote control, or operate the cooler lid, and possibly both at the same time.

WHAT ARE FRIENDS FOR?

Cowboys aren't especially known for not ever doing anything stupid. In fact, they are fairly consistent in that department. That is proven apparent in four out of five stories they tell you.

Usually "stupid" is directly related to something they shouldn't have roped, something they shouldn't have ridden or, sometimes, someone they shouldn't have married.

Right after that list comes the list of pranks they pull on each other that should have, but didn't, get them killed.

Another constant, but not always a requirement, is the influence of an adult beverage of one variety or another.

Johnny and Tom were serious ropers—or as serious about roping as they got about anything. They were wheat pasture punchers by trade. This particular winter they had cattle scattered all over the Panhandle.

They had been doctoring sick cattle all day, roping and giving shots. Late in the afternoon, they decided it would be prudent to test a little of the fresh moonshine they had recently acquired.

They unsaddled at the pens and were making a plan for the next day while conducting this experiment on the 'shine.

The pens were home to a bunch of chronically sick calves and after a while, they decided that these needed to be roped. No reason needed, except they were there.

The kids' bicycles were leaning up against the fence and, well, transportation is transportation.

The bicycles worked fine; these calves were pretty sick.

When they got all the calves roped, standing in the pen, there was a big Santa Gertrudis stray cow that belonged to a neighbor.

This old hussy weighed at least 1,200 pounds, with horns and was about half-wild.

Tom bet Johnny he couldn't rope the old cow and, of course, Johnny couldn't pass up a cowboy challenge.

Johnny shook out a loop and snared the hussy right around the horns with the other end of his rope tied off to the handlebars.

He just couldn't help himself or pass up this opportunity. Tom opened the corral gate to the outside world. The cow left on a run with Johnny trolling behind at the end of the catch rope.

Johnny was holding his own with the bicycle until it hit a hole in the pasture. He had on leggings and spurs and as he dug in to hold on, the spurs shelled all the spokes off the back wheel of the bicycle. In the inevitable wreck, the old cow drug the bicycle off, leaving Johnny laying in a heap on the ground.

Their wives were supposed to come out to the pens that evening and bring them supper. The cow had run through a gate tearing it up and rescuing what was left of the bicycle didn't make anybody happy.

Supper that evening was a very solemn affair. Wasn't a funeral, but may as well have been.

NEW TWISTER INITIATION

Nothing smells better to a cowboy than a sweaty horse as long as it's his favorite. Well, except maybe a pretty girl, as long as she's his favorite.

A quality horse worthy of a cowboy's love and respect doesn't just happen– there are many hours and countless miles invested between a pretty little colt and a good finished horse.

Starting colts is part of the normal operation on most ranches. The young cowboys look forward to it and a new crop of colts will keep them busy.

Sometimes extra hands are hired to help and occasionally there's a twister who likes nothing better than to ride a horse that likes to buck. The horses will get in the spirit of this and some will pitch every time a rider swings up just because both the rider and the horse like to show off.

Billy was such a horse. With this reputation, he was known as the "initiation horse," saved for the new hires.

Initiations provide a little fun with newcomers to see if they were worth their salt. They would rope Billy out of the horse herd, and he would stand quietly while he was saddled and the cowboy got topside.

Once the new twister was aboard, he would swallow his head. If the twister made a ride, he was accepted as an equal. If he was thrown high enough for the birds to build a nest in his hat, times would be hard for him for a while.

If a new hand didn't strike the old hands just right, they would skip the initiation "for fun" part and Billy would be on a temporary vacation.

A new twister had shown up at the ranch late one evening. He allowed that he had come to help these boys out a little.

He had recently been working at the feed store in town, but was ready to out-punch anybody around– assured that he was loaded up with cowboy skill.

The regular cowboy crew looked him over, took in the boots with 18" high tops, under-slung heels and britches tucked in. They saw that he had a hat with a big turkey feather. They didn't miss the attitude, either.

Before light the next morning, all the hands gathered at the horse pen. The wagon boss was roping out horses to work that day. The new hand stood around, anxious to get to work on this big, prestigious outfit.

With dead accuracy, the boss laid a houlihan loop over one horse after another for the waiting cowboys. When the crew had their mounts, the boss dropped a loop over a big, stout-looking dun.

When the loop settled, the dun set back, blew a few rollers out of his nose and wouldn't come out of the herd. Finally, one of the hands, who was already mounted, had to dally him up and drag him out.

The boss told the new hand: "Here's your horse. His name is Sam Bass, he's 7 years old and you won't need that bridle. He's still in a hackamore."

Proving that attitude doesn't always replace intelligence, the new twister took in all the expectant faces, looked over the dun blowing snot in front of him and told the foreman, "You can take Mr. Sam Bass and stick him up your ..."

It rhymed. No one knew that this new twister was also a poet.

The turkey feather pluming above his hat was last seen fading off into the sunrise as the feed store cowboy headed back to town.

THEY DON'T ALL STAY AT THE RANCH

Ranch raisin' is the best way any kid could grow up. It teaches life's lessons in the most basic ways for all the formative years. Then one day the ranch kid wakes up and he is an adult with big-people dreams and plans that often don't include staying at the ranch.

Cowboy life is not always awash in glamour. A good bit of it involves farm implements, a shovel, plumbing tools, fence stretchers and carpenter skills. It's that seedy side of the career than will send a young man looking for the excitement.

My own younger brother was a good example of this natural migration. Being the last one left at home – his older brothers already gone, making the big bucks in the construction industry – he was the last cowboy standing. Well into that stage of maturity where Dad didn't know a thing about anything, ranch work was just another frontier of disagreement. To improve his circumstances, he joined the U.S. Army.

The irony was that after basic training, they sent him to live on an Army-owned ranch in Colorado to be in the cavalry regiment that performed in parades and re-enactment. In this newfound career, he cleaned stalls, tended to horses and other regular ranch chores.

Eventually his tour was up and he moved back home. However, cowboy glamour at the ranch soon left again in the midst of moving sprinkler pipe in the hay fields and other such tedious chores.

The lad re-enlisted, but not without throwing in a tour of duty tending bar in a big city honky-tonk, teaching country dance lessons and landing a role in a beer commercial for the really big bucks. His cowboy expertise came in quite handy.

Through the years, the armed forces have been the "great escape" for a number of cowboys. In Texas, Donnie was still at home, helping his dad out with the family ranch while considering his career opportunities. During the last rainy spell, his dad decided that the house roof needed new shingles. Donnie came to mind for this job.

Soon, Donnie found himself up on the ridge row with a shingling hatchet, a bundle of shingles and a nail apron firmly in place. After the first couple of rows of shingles, the summer sunshine was losing its charm.

Donnie managed to work the head off the hatchet and was climbing down when Dad, who was paying more attention than Donnie realized, tossed him up a hammer. It took Donnie a while to break the handle on the hammer, but he got it done. Dad had another one ready for him.

While nailing on a couple more rows of shingles, he decided that Dad was preoccupied. The hammer slipped out of his grasp and landed about 500 feet away in the stock tank. That just happened to be the last hammer on the place, so Dad sent Donnie to the hardware store in town for more hammers to finish the job.

On the way to town, Donnie noticed he had a couple of extra shirts and pairs of Wranglers in the truck so he went on down the road to Mineral Wells and joined the Army. Three years went by before Donnie returned stateside and got back to Texas where he headed home. On the way, he remembered what started it all.

He stopped by the hardware store, got a couple of hammers and when he walked in the door at home told his Dad, "Here's your hammers. I got them just like you told me."

COWBOYS AND POT-BELLIED PIGS ARE A BAD MIX

I bet you thought this was going to be a New Year's Eve party report.

With that same title, it could be, but it's not. It really is about pigs, real pigs, and in this case, a couple of the pet pot-bellied variety.

Good cow horses will stand for a lot of things that would make the run-of-mill-backyard-variety equine lose their mind, jump upside down and get you hurt.

Rock had calmly avoided rattlesnakes, pheasants flying under his belly, bulls charging him and deer blowing out of the brush nearly on top of him.

He even put up with pilgrims in pink spandex pants petting him and children rolling under his feet. However, a pot-bellied pig did him in.

New to the area, the cowgirl and her cowboy had gone to help some neighbors gather cattle. As one does when one is at someone else's outfit, she got her instructions and began trailing cattle back to headquarters.

Following a little bunch of cattle that was headed to the main herd, she and Rock seemed to have things under control until out of nowhere a flock of sheep came on the run right through the middle of the cattle.

Rock pinned his ears and squatted back on his hindquarters but held his own against the white, hopping invaders without ever swapping ends and making a run for it. The worst was yet to come.

After the cattle had been re-gathered with the requisite amount of cussing, the corral sorting work was done and the waiting semi-trucks were loaded.

The horses were tied to the corral fence when the crew sauntered toward the house for lunch.

About halfway through the meal, the cowboys noted all their horses were in a dead run out across the pasture.

The first job was to capture them and then try to figure out what had set them off.

Next to the fence where all the broken bridle reins were hanging sat a fat, happy, very ugly pot-bellied pig.

Horses snorted and shied from the fence and the pig looked them over inquisitively, hoping for a little more action out of them.

In this same part of the world, a rancher woke one morning to silence.

Not a good sign when he had, just the day before, weaned five pens full of calves that should be bawling their heads off for their mamas.

He raced to the corrals to find all but one pen of calves completely gone. Adios, por allá, missing! Gates torn down, fences laid over.

What he also found was a pair of pot-bellied pigs that had wandered a few miles to create such havoc. They'd taken up residence and seemed to think they were right where they belonged.

Those hogs left a lasting impression on the cattle, the horses and the cowboy.

He returned the pets to their home but once they had discovered the trail to so much fun, they made the trip often. Kind of like relatives that show up uninvited, stay too long and don't know they are not welcome.

For years after the event, the cowboy's horses snorted, shied and acted like there were unseen monsters in that set of corrals.

One of the pigs relocated when their owner did while one of them went MIA and is now listed with the rancher "cold-case" files.

The moral of the story goes with the saying that "good fences make good neighbors." It's a dangerous world out there. Keep your pigs at home.

AND THE HORSE HE RODE IN ON

Cowboys that hang in the same circles identify each other by their ride long before they recognize a face.

In the world outside of cowboy, if you were to inquire about someone in town, you might get a litany of descriptions. "Gwen? Sure, I know her. She's a marine biologist, has a couple kids and lives over on the hill above the golf course."

Or, "Bob? He's looking pretty good for his age, but he did just win a Nobel prize for something no one understands. He's some sort of nuclear physicist. He lives down by the lake, has a boat at the marina."

With cowboys, the dialogue will go something like this:

"Dan? He's the one that rides that well-made paint horse and ropes heels. That's sure a nice horse. I used to have one about like that and, man, I won a lot of money on him."

Or, "Jess? I don't know him personally, but I sure like that big blue horse he rides."

Cowboys will notice and evaluate a horse long before they even look at the rider. When they do get around to noticing the rider, they will already have an impression of the type of person who would ride a horse like that.

The stars of the rodeo world are no exception to this ironclad rule.

Cowboys who are likely to never come into personal contact with the sport's champions will be very aware of the horses they ride.

Ask any cowboy that's paying attention what kind of horse Clay Cooper rides, and he'll be able to tell you a good bit about the favorite gray horse.

They will all be up on the fact that nine-time National Finals qualifier Stran Smith has a new mare named Destiny, to replace his long-time ride, Topper, that was killed in a hit-and-run when he wandered onto a highway.

Some horses achieve as much fame as the riders. Viper, Speed Williams' good horse, has a rope named for him. Many pastures, barns, ranches, sometimes even the children, are named for a favorite horse.

The women of the sport are no exception. Every barrel racer in the world can tell you that Charmayne James spent $150,000 getting her world-beater, one-of-a-kind horse, Scamper, cloned.

In the world of ropers, cowboys are often introduced according to their horses. They might be marine biologists or nuclear physicists, but no cowboy will ever know or care.

Jess was calling around lining up new partners for a big upcoming roping. He had gotten a phone number from another roper by describing the horse ridden by the man he wanted to call.

When he called the man, the first thing said after Jess introduced himself was, "You ride that big blue horse, the one with the brand that's a bunch of numbers?" Everybody was identified, so the deal was made.

Cowboys will spend more time getting their horses ready, tuned-up, tack checked, trailer ready, gear, feed and medicine loaded, than they will on themselves. Hours will be devoted to the horses, generally starting way ahead of any scheduled roping or rodeo.

About fifteen minutes before time to leave, they will run in the house and grab the first shirt in the closet. Good to go.

They all know that how they look doesn't much matter to any other cowboy, but they will be judged by the horse they ride in on.

YOUNG BUTTONS

There is very little more exciting to a youngster at the ranch than branding time. It represents a coming of age for them as they work their way up the ranks through skill-appropriate jobs in the branding pen.

Spencer was 11 and had been to every branding since he was big enough to walk. However, this year was different. This was the first time he'd be allowed to spend the night in camp down on the river with the other hands.

As a special treat Jim, the boss, allowed Spencer to invite two friends who looked to be about the right size for flanking calves. One was another ranch kid, the other a town kid who aspired to one day be a cowboy.

Allen, the camp cook, had gathered up plenty of firewood, good groceries and all that he needed to feed the crew supper and a big breakfast.

After supper, the requisite campfire tales were told while the boys listened wide-eyed. When the adults drifted off to sleep, they decided to do a little exploring, run a bit and play awhile. This required keeping the campfire going.

This went on well into the night and until wee hours of the morning. The excited young "buttons" decided there was no need to sleep now, so they just kept the fire going while they swapped more tales, waiting for their big day to get underway.

When Jim came down to join the hands for breakfast, the cook was mad, which is never a good thing under any circumstances. Everybody was standing around with saddled horses, but nothing was happening about breakfast. When Allen calmed down enough to speak, he reported there was no firewood so there was no food.

Telling him it was not a problem, Jim said they would go gather the first pasture to give him some time and then be back later to eat.

Jim instructed the boys to go with him and they followed along thinking this was their big opening to go with the boss. Jim was riding a colt that needed some miles, so he took the outside circle, dropping off cowboys along the way. Three sleepy young punchers continue to follow behind the boss.

When they got to the pens with the cattle, everybody ate and the work started. The boys were assigned to the flanking crew as two ropers drug calves to the branding fire in a rapid procession. The young buttons didn't have any time to think about much else except the next calf coming at them.

By noon the first pasture was worked, momma cows and babies paired back up and turned out. Jim sent everybody except the boys to the cook's wagon to eat. The boys were told to gather firewood, then eat last. Just as they hit the wagon to eat, Jim was ready to gather the second pasture and said, "You boys come with me."

The second pasture was the same routine. The boys got to perfect their flanking and another chance to gather more firewood.

Late in the afternoon after the third pasture was gathered, worked, flanked and a little more wood gathered, the neighbors and hands that had come to help drifted off toward home.

Jim told the boys to gather just a little more firewood in case they needed to brand again some day. Lesson learned, Spencer knew better than to object and the other ranch kid was in robot mode. The city kid decided to become a lawyer.

The good news is that for the next 20 years, no one would have to gather any more firewood to cook breakfast.

Cowboying is tough work – being a "young button" can sometimes be tougher.

CASEY AND THE GRULLA

Cowboys are quite often loners, especially if they are holding down a ranch job in a remote area of the West.

Casey was just such a loner, with his horses serving as his best friends.

For five years running, he wintered very peacefully at the Box Canyon Camp on the Estacado Comida Ranch in northern New Mexico, a job that suited him perfectly.

With an abundance of horse charm and the fact that the ranch raised premier colts from their large mare herd, it was a good match for them all.

The mares were wintered in the large, protected canyon, giving the colts that were born in the spring a safe place to grow and a chance to learn to travel in the rough country.

Casey would work a little with the young ones, teaching them a few manners and getting them used to being around that beast called man. It was called work, but for him, it was something next to heaven.

He kept the mares gentled down and, when needed, offered a little midwifery skill during foaling.

Then along came the ranch owner's son, fresh out of college with a degree and an assignment to reacquaint Casey with reality.

Brad arrived at winter camp full of book ideas, enthusiasm for the invigorating outdoors and thoroughly in love with a blonde who promised to wait for him until spring.

During the cold months, Brad got over his scholarly schemes and recognized the environment for its greater challenges, but when it was time to bring the mares and colts out of the canyon, he was still in love.

Casey usually hit the rodeo road during the summer. Brad decided he and Blondie, who happened to be a barrel racer, would summer with Casey on the rodeo trail.

He was, after all, the boss' kid, so off they went to collect the girl and get on with the summer.

When they met her at the arena, both men's eyes lit up.

This was a very pretty girl – long honey blonde hair and big green eyes. She was riding a big grulla gelding built like a fine quarter horse should be.

Brad had his eyes on the girl, but Casey had eyes on the gelding.

Through the summer, the cowboys kept their pockets lined with their rodeo winnings.

However, Blondie wasn't faring as well with the grulla, who hadn't quite caught on to the concept of running around three cans.

Casey regularly applied his "horse charm" to keep the big slate-colored horse calm and working well enough to get Blondie to the next rodeo.

He was not charming enough to keep her consistently winning, not that she noticed.

She and Brad spent more time looking at each other than at the rodeo schedule, so it didn't appear to be a career goal for either of them.

When the winds of autumn began cooling the days and fall nights took on a crispness recognized as a precursor to winter, Casey began to think about Box Canyon.

He had one more thing he wanted to accomplish before summer's end.

Brad and the honey-blonde were making their own plans and a fall wedding was in the works.

Casey agreed to stand in as best man at ranch headquarters along with further arrangements to leave right after the ceremony.

He and the newlyweds had come to an agreement about the future.

After the last toast for lifelong bliss was made, Casey changed into his jeans, a Carhartt jacket and set off for winter camp.

Brad had the girl. Casey left riding his gray and leading the grulla.

And everybody lived happily ever after.

Now, don't you feel all warm and fuzzy?

FOREMAN OF THE ROUGH RANCH

He wore a hat that should have had a burial five years ago, but the turkey feather stuck in the hat band was in fair shape.

The lean, wiry puncher, with his britches in his boots, was a reputation kind of cowboy – reputed for his abilities to handle cattle and horses – and for doing things "Darrell's way."

Everybody that worked around Darrell knew there was the easy way, the hard way and Darrell's way. Darrell's way made the hard way look easy. Legend has it, Darrell wasn't afraid to make his point, if needed, with his fists.

One time Darrell had the neighbors gathered up to put a couple truckloads of fresh cattle out on wheat. The first job was to get the calves to stop running after they bailed from the back of the truck. Finally, the critters were convinced they were surrounded with Darrell riding point, a few hands on both sides and a particularly aggravating cowboy riding drag at the back.

This cowboy, out of pure meanness, kept pushing the cattle up over the top of Darrell in the front in an attempt to make the cattle start running again. Darrell would ride back and give him the "what for" and then the cowboy's wife would bail into the argument. While the "discussion" often ended in a fistfight, everyone else had to stop the cattle from running off.

For all Darrell's poor ways with handling some of his help, he was better than a good hand with cattle, had a strong work ethic and made an excellent ranch foreman. While he probably could have had his choice of ranch jobs, the one that suited him best was on a ranch with rugged, rough terrain.

You see, Darrell had a weakness. He liked to ride colts. He liked it even better if they would buck before he could get them gentled down. Horse whisperer techniques didn't interest him. His training methods included lots of miles and slow, quiet work around cattle. Darrell could get more work done on a colt than most seasoned hands on a broke horse.

One year, late in the fall, everyone who was friends enough to come help Darrell ship yearlings gathered at the Rough Ranch. It was threatening to snow but after breakfast, still in the dark, he got the entire crew horseback and headed for the backside of the outfit in a long trot.

Once there, he began dropping off hands, telling them to gather the cattle to the hilltop in the middle of the three-section brush-covered pasture. The hands knew the weather would likely catch them but also knew these cattle would cross the scales for a payday when they hit the pens, so needed to be handled slow.

The cowboys straggled to the hill top with small bunches of cattle, everyone accounted for. Darrell strung the cattle out, counted them and then counted them again.

Then he told half the crew to hold the herd and sent the rest back to the pasture to pick up strays. The weather continued to threaten and after a time, the cowboys all came back but no one had found any more cattle.

They drove the herd to the pens, weighed and put them on the waiting trucks. Darrell reported to the Rough Ranch owner that every head had been accounted for.

Later in the day, someone asked Darrell why he had sent the crew looking for strays if he already had the right count. He didn't answer, but anyone that knew Darrell knew what the deal was. He simply wanted to put a few more miles on his colt before the day was over.

There is the easy way, the hard way and Darrell's way.

COWBOY LORE FALLS TO NEW LOW

With the stealth of a Ninja fighter, the cowboy eased his way around the end of a 20-foot stock trailer, hunkering his tall frame down far enough to stay out of sight of his prey.

With deadly precision he, in the flash of time it took for a single thought, slammed the gate closed and his job was done. The last turkey hen was loaded.

There was to be a June wedding in the yard at the ranch. That same yard also happened to be home to a flock of wild turkeys, a few of which had been relocated there some years back. Now their numbers were tripled.

These big birds roosted in the cottonwoods, perched on the vehicles, decimated the flower garden and left unpleasant reminders of their recent presence.

So a turkey relocation program was "hatched" by the head cowboy.

This same man plans a major cattle working in a matter of hours but this project would take at least two weeks. With careful cunning, he began baiting the turkeys into the trailer by trailing feed down the length of it.

When the time came that he could get a few captured, which sounds easier than it was because as soon as they'd see him they'd fly out, he'd shut the gate and haul them to a grove of cottonwoods at the south end of the ranch.

This took three trips for 14 turkeys. The last trip was for a lone rebel bird who refused to be captured, inspiring a new level of a stalking-capture mode.

I missed the photo opportunity of the year – a cowboy hauling one turkey in a 20-foot goose-neck trailer.

While it truly needed to be done, the very idea of it takes the cowboy image to a new low.

On the upside, it certainly has been fodder for moments of hilarity as the tale has been told and retold.

During a recent discussion of the turkey-herding incident, it was mentioned the turkeys had returned to their first home one night last week. The return just happened to coincide with the arrival of a new grandchild whose parents also reside at the ranch.

While the incident could seem somewhat mystical and the oh's and ah's momentarily sustained the coincidence, the reality was hard and cold. It was pointed out these were notoriously dumb drown-in-a-rainstorm turkeys – not baby-delivering storks.

In looking for a, perhaps, positive use for the turkeys besides Thanksgiving dinner and turkey sandwiches, it was suggested that they be painted white. And if a process of launching them could be engineered, they then could be used at the wedding instead of white doves.

The suggestion brought a look on the bride-to-be's face that could only be interpreted to mean this wouldn't happen in her lifetime.

Another response to the jovial turkey herding story came from an Albuquerque friend of the turkey herder. He wrote:

The Gobblers Shuda Went to Town

Darn bro...

I heard u was a turkey man.

A turkey man what am!

U shuda brung dem

turkeys to 'querque

We'd a put'um in a pot

an eat'um onda spot.

Yup...them turkey's uda

stayed right here in 'querque

OK, it's not Whitman or Emerson, but it is funny, all things considered.

The next story I'm waiting for is the response of the saddle horses when they are asked to get in that same trailer. Horses are funny about loading up in trailers that have hauled anything other than a horse or a cow.

Try loading one after a hog hauling.

REJECTED BY eHARMONY DOT COM

Never did I think I'd know anybody personally that would sign up for a "perfect match" on eHarmony, let alone have the outfit tell them, "Sorry, no can do." However, the news of the rejection was made public this week.

After much sport at the expense of the rejected lad, I pondered the situation further.

I'm not sure what kind of questions an online dating service asks, but if truth is forthcoming, I suppose there are a number of folks with traits that make them unmatchable.

Cowboys, as a rule, most definitely have some traits that make them charming characters but not always desirable as a keeper.

While having skills that mark them "cowboy," they often don't have much finesse in other trades.

For example, there is Billy.

Billy had gone to the Texas Panhandle to work in one of the feedlots. As it does in that part of country, the winter got much too cold for him (and anybody else in their right mind to be horseback riding cattle pens.)

He recalled his grandmother's fondness for warmth and that she liked it in the old folk's home where they keep it about 85 degrees all day.

So he went to the local old folks home and got a job as the maintenance man. He didn't know anything about maintenance beyond how to fix floats on cattle waters but that didn't slow him down.

There were a few incidents, but the old folk's home didn't actually fire him until somebody flushed a toilet and the lights went out. Billy is now back home in south Texas.

A prospective date might scowl at the fact that Billy's rope horse, Hombre, is his drinking buddy. Hombre is about 30 years old but Billy figures he only has to run about five seconds at a time so he continues to use him in the roping arena. Hombre can hold his liquor a little better than Billy but it might simply be that relative to his size, he doesn't get quite the load.

The recent eHarmony reject received a follow up letter from eHarmony explaining some of their issues for dismissal. In part it read:

"A guy in a skirt is not attractive, even if you call it a kilt in reference to whatever it is you wanna-be Scottish boys wear when you want a breeze around parts usually covered by Wrangler.

"The picture of you with the potato in your pants was a new wrinkle for us. We weren't sure what you intended by the location of said spud, but we advise that in the front instead of the back would provide a more suggestive appearance.

"The pickup line, 'Hey Baby! Want some of this?' went out years ago. In fact, it was never in.

"In reference to your last date, there is the reminder that if you holler 'put out or walk,' you should be in your own vehicle.

"We do think there is someone out there for you. Our suggestion is a mail-order bride from Russia with a mustache who has never met you prior to the nuptials.

"Thank you so much for the application. We will frame it and keep it as a reminder that we were wrong. We can't help everyone."

I'm thinking that most eHarmony rejects will resort to the old-fashioned way of picking up girls – buying them a beer at last call just before closing time.

THE WALL STREET COWBOY

Cowboys are philosophers. Even Gus and Woodrow of *Lonesome Dove* fame were known to "talk a little philosophy" from time to time.

Maybe it's the solitary work days and time they have to think things over. The more serious ones are planners. NASA can get a space shuttle launched with less planning than many ranchers put forth before a big spring branding.

When Rocky's dad called him home from the feedlot, the news was bad and good. The bad news floored Rocky. Dad had sold the ranch. The good news had the same affect – Rocky now had a pile of money.

Rocky understood almost all aspects of the cattle business but that association didn't often leave him with many dollars to tend. He decided to talk with his good friend, Tex, who was also sometimes his roping partner, a top hand, good cattle manager, good money manager and an ace horse trader.

Tex took a deep seat on the porch and prepared to share his philosophy on money.

"You want to know what an asset manager is?" Tex began. "Remember when we were rodeoing and Wes always held our entry fees and gas money so we didn't spend it foolishly? Wes was our asset manager."

"What's a mutual fund?" continued Tex. "You remember when you, Wes, and I partnered on that bunch of colts? We each had a set to break then we were going to sell them all together. We knew that if one colt turned out to be an outlaw and we lost money on him, likely one of the others would be a little extra special and would sell for more than average and it would balance out the loss. That's a mutual fund."

Tex recalled that Wes again handled the money – dealing with the bank and brokering the sale of the colts. "That made him the fund manager," he said. "He got a little extra off the top for handling the paperwork."

"And a bond," explained Tex, "ain't nothing more than a loan. I could loan you some money, you would give me an I.O.U. that said you would pay me back at some certain time, plus interest. You would pay me a little bit along the way so I could keep up with my bills and then pay it all at the end of our deal."

Tex gave Rocky a run down of annuities, equities and a number of other financial terms and related it all back to horse trading. He and Rocky decided that Wall Street people must have, at some point, been horse traders, too, and just fancied up their vocabulary for the city folks.

After Rocky felt a little more comfortable about handling his new-found wealth, he and Tex settled back to ponder thoughts about why his dad had sold the outfit. Although the ranch has always supplied a steady income, some years better than others, it more dependably supplied a steady supply of work.

With his new-found money-planning knowledge, Rocky could already see that the profit margins on the investments were going to out-distance his former ranch income by far. That realization somewhat soothed the sadness for the ever-changing landscape of family ranching.

Besides, Rocky was pretty sure Wall Street could always use another horse trader – one with a direct look and the firm handshake of a cowboy.

HORSES AND SNAKES TOP THE STORY LIST

There are some very fine stories that never make the history books.

Storytellers keep Native Americans' legends through the generations. It's the job of a gifted tribe member to be the keeper of the stories and to pass them on to the next generation from the many generations that came before.

Cowboys do much the same thing. The Native American storyteller will have a name like Grandmother Two Bears or Old Father Story Teller, the cowboy will simply be named Ben, Joe or Charlie. But if they were to be in a tribe somewhere, they might be named something like Man with Crooked Legs or Hitch In Get Along.

Old cowboys tend to be shorter than they were in their youth, a bit bow-legged and they waddle when they walk because their parts don't move like they should. The days of the long-legged strolling stride left when arthritis arrived.

What they no longer have in athletic ability, they have maintained with humor and the passing of the legends, otherwise known as cowboy wild and woolly tales.

The number of topics from the old days when cowboys were king is endless.

Always, things were bigger, better and wilder "back then." They may not be able to give you their wife's full name accurately and certainly not the date of her birth, but they recall every single ill-headed horse they ever rode in a 50-year period.

Guaranteed, they can call by name every horse in every story of every wild wreck they ever had that involved a rope and cow.

And for some reason, each horse will either be the best he ever rode, or the sorriest. Recall has managed to sort out the possibility of any mediocre saddle horses from days gone by.

Another topic that will bring on the windy stories is snakes. There are generations of big ugly diamondbacks that slithered into a bedroll, traveled up a catch rope to meet the roper or fell out of a tree on an unsuspecting cowboy.

Snakes, in their mystical ability to strike fear in the hearts of all men, garner a corner of cowboy history all by themselves. Ask any bowlegged, cowboy-booted hombre you run into for his best snake story. I promise you he'll have at least one.

Then there are the "goin' to town" stories. Not so long ago cowboys went to town only occasionally and that trip involved buying some groceries and other supplies.

On that same sojourn, they might eat a steak at the local restaurant, spring a few bucks for a haircut and spend some time at the local watering hole imbibing in some cool spring-water beverages.

One of my favorite cowboy storytellers had a great tale that involved both a trip to town and a horse.

During one of those supply trips to town, he decided a cool one was in order. No sense tying up outside; he just rode the green-broke colt into the bar.

Things were going seemingly well until, suddenly, the jukebox music stopped. When the place got quiet, the horse got wide-eyed and made every attempt to get to somewhere else.

In doing so, he fell over onto the pool table breaking his rider's foot. For decades, the cowboy blamed the jukebox for his injury that kept him gimpy the rest of his life.

Cowboys in the last half of the last century lived a kind of cowboy way that, for the most part, has faded from the landscape.

The new West is continually inspired by their stories.

COWBOYS AND ATVS,
AN UNNATURAL MIX

For 150 years, cowboys have ridden horses. It didn't always turn out well, but generally it has been the perfect combination for the job.

Then along came the motorized "horse" in various versions of wheel count – two, three and four.

Dirt bikes, the rugged slower model of motorcycle, made a stab at offering an easy "saddle up" for the cowboy along with the promise of covering lots of country in a short amount of time.

Hunkered over his iron steed, hat pulled down tight, the cowboy did indeed find he could take the outside circle, bring the cattle at a high trot from the farthest corners and never have to let his "horse" have a breather.

If he could still walk at sundown, he might make mention how that bike like to have beat him to death on the rough terrain.

But as is his way, he would cowboy-up and do it again the next day.

Like with great bronc stories, this screaming machine created many a tale.

"Man, you shoulda seen it. I was watching ole Roy headed down the top of the ridge when all of a sudden he just disappeared! He flat just dropped out of sight."

The storyteller, who happened to be proudly mounted on a real horse, would then tell about loping over to find Roy and his iron mount at the bottom of a sink hole. Roy and "Trigger" were in need of some serious help that would come only with a rope and a pull from a real horse.

Then along came the three-wheeled ATV followed by the somewhat more stable four-wheeler.

There are cowboys in cowboy country that will tell you, "Give a cowboy an ATV and in a year, you have a dead cowboy." It sounds harsh, but the phrase was born from a sad truth.

For every comical tale about the new generation of motorized cow herders, there are twice as many proving the danger of the machines.

While accidents happen and accidents are accidents, there is something menacing about mixing the wild-side nature of the cowboy with something mechanical.

They truly cannot resist pushing the limits, testing the parameters and making every effort to prove something that never needed proven.

If fast is good, faster is better. If gradual is safe, straight up or down and all out is better. "Hey Maw, watch me rope that sick yearling, dally to the tool box and turn off hard enough to bring him to Jesus."

Convenient, step-saver, fuel-saver, time-saver and man's best tool when conditions warrant it, the ATV has unequivocally become standard equipment for ranches and farms.

There isn't anything better for the job when you need it. Handy as a pocket on a shirt and likely to be around just as long. But still very dangerous.

I have a teen cowboy in the house now sporting a neon-orange leg cast. His recipe for a wreck that could have been oh-so-much worse was a rope and four-wheeler.

He was headed out to rope something – the rope coils thrown over his shoulder while he moseyed the four-wheeler to the corral. The tail of the rope got hung up in the rear axle and in a flash, the rope cinched down tight around the lad's shoulder and arm. He was jerked off the seat, slammed to the ground and drug a short way until the machine came to a stop.

There was a lot that could have happened but thanks only to God, it didn't. He will be uncomfortable, inconvenienced and no doubt a little bored for the next few weeks. But I'm very glad he's able to experience it all. Maybe he'll come out the other end a little wiser.

Can 13-year-olds be wiser?

THE JINGLE OF THE SPUR

There are several items of a cowboy's life that he treasures highly, sometimes more than life itself.

Right at the top of the list are his spurs.

Spurs are a standard part of a cowboy's day-to-day working equipment and, usually, they never come off his boots.

Understanding that cowboys seldom remove their spurs will dictate the type of furniture, carpet or other flooring in the house he lives in.

It is also a reassuring sound late at night when the work has gone long past sundown, to hear the familiar jingle come through the back door.

Spur jingles are as recognizable to a ranch wife as the sound of the ranch pickup to the resident dog.

Spurs are a point of pride for the cowboy. They regularly are handmade, inlaid with silver, embossed with initials, brands and good luck totems.

A spur can reveal a cowboy's geographic origin, riding philosophy, ego or intelligence quotient.

Foot jewelry will include bells, jingle bobs, pizza cutter rowels, tooled straps, and silver buckles. Whatever they look like, it can be, and often is, a topic of regular conversation.

By nature and more often by necessity, cowboys are frugal. All sorts of items will be recycled until safety becomes an issue and even then it is a debatable moment.

Case in point, one cowboy I know owns a tie-down strap, homemade from material he and his buddy partnered on when they were in college 30 years ago.

That strap, a mark of frugality, ties to a new $3,000 saddle set on a very expensive horse to go to some high-dollar ropings.

But he sees no reason to replace it as long as it still does the job.

He is sure that strap is the reason his horse can stay on course when he's after a triple-A running steer.

This same frugal cowboy shares the story about another cowboy, a very economy-minded individual, who needed to replace his worn-out spur rowels.

In his mind, the store-bought rowels were simply too expensive. So he decided quarters would work just as well. He could put them in himself and avoid the saddle-shop labor component.

The next time these two models of frugality met, the former 25-cent spur-roweled cowboy was wearing nickels for spur rowels. He said he had saved 40-cents that way and the horse didn't know any different.

The rest of story was that it took a week for him to figure out how much he'd save before he could determine if it was worthwhile.

Bill, a Texas Panhandle short-grass rancher, had a cowhand working for him who was especially proud of his spurs. The cowboy wore them all day, every day and even presumably at night for a number of years while working for Bill.

At one point, the cowboy needed a loan. He needed $600 for some undisclosed emergency and to prove his sincerity, he offered to put up his beloved spurs for collateral.

Bill felt like nothing short of threats of life and bodily harm would part that man from his spurs, so he loaned him the $600.

Whatever the emergency was, the cowhand skipped the country with Bill's money.

The spurs are now a family heirloom, known to all the extended cousins down the line as "the $600 spurs."

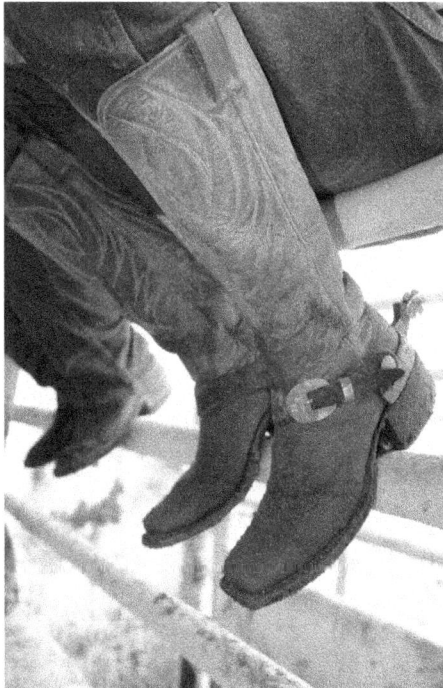

DAN THE TEAM ROPER

A TALE OF SLATS, DAN AND WHITEY

Slats is a team roping horse worthy of a story. His owner Dan follows narrowly in second place for his story worthiness. Then there is Whitey the dog. These are true stories with the names unchanged to protect no one.

From a regular person's perspective, Dan looks to be close to 7 feet tall, weighing in at about 80 pounds. Both numbers are obvious exaggerations but you now have the mental image of this very tall, very narrow cowboy.

Perhaps the only 17-hands-high heeling horse south of the Mason-Dixon line, Slats is a trim 900 pounds. His height is not a handicap, only because Dan's arms are long enough to compensate. Slats seems only slightly self-conscious of his height as he stands head and shoulders above the other heeling horses when tied to the fence during the beer-drinking breaks at the arena.

This fine animal's resume begins with his career as a race hose in El Paso where he received all the appropriate training and made two starts. According to Dan, Slats was so polite by nature that when he saw all those other horses in such a hurry, he just let them go on by. He came in dead last both times.

Management moved him to new track-related position as a pony horse where Slats remained for a number of years. Sadly, he became a victim of company downsizing, was traded and was soon on unemployment. In this case, that also meant a starvation diet.

Dan was day-working for the man that owned Slats, saw his plight, and asked if he could have him. The deal made was that if Dan would pay the vet bill to get Slats back up to using shape, he could have him. To date Dan has $382.50 in Slats plus another couple million dollars in wormer and feed.

Slats even has a pet. When Slats first came to live with Dan, he was thin and weak and, according to Dan, thinking about dying. Dan thought a buddy might make Slats more interested in continuing life and making a career change. Whitey the dog became Slats' full-time partner.

When Whitey first arrived to live with Slats, he spent a good portion of his time chasing him. Now that Slats is back in good form, he chases Whitey. Dan's description of Whitey includes "he is about so-high, has a medium length basic black coat, trimmed in black."

On occasion, Dan offers Slats to his best friend and fellow team roper. Tim is of average build, 6 foot tall or so, and when Dan says he can use Slats to make a run he always gives the offer some consideration.

Looking at Dan and then at Slats, Tim says, "It'll take me a month to reach your stirrups." Dan volunteers to take up the stirrups about four holes.

Tim backs in the heeling box. Slats does his usual rocket launch break which Tim is able to ride him through. Quickly in position behind the steer, he casts his loop. The minute it leaves his hand, Slats buries his front feet in the ground and the back pockets of Tim's Wranglers are precariously day-lighted from the saddle.

Back at the roping boxes, Dan laughs and says to those standing there, "I love making him ride old Slats. He falls for it every time."

That kind of friend is hard to find.

KID IN THE FARMER BROWN HAT

Cowboys are generally good ol' boys but they are also fairly set in their notions about certain things. They pride themselves in being able to read cattle and read each other when it comes to quality and skill.

Dan had seen twelve summers and was spending his school vacation working for his grandpa on the Matador Ranch in northwest Texas. His grandpa was a top hand and had worked on the historic, sprawling, short-grass ranch since he was a kid himself.

Grandpa had two sons who were "going down the road" trying to make their living in the rodeo arena. They were good enough to win their entry fee money back and a fair amount of walking around money with fast times in the calf roping events around Texas.

These pretty-fair ropers were Dan's uncles and they were coming home July 4 to enter the big calf roping at Post.

Dan was doing a man's work and had a man's string of horses assigned to him. However, since he was the last "man" hired, he hadn't gotten the best pick of ranch horses from the herd.

In his string was a big, tall, rawboned gelding –heavy-made indicating some Percheron blood somewhere not too far back in his lineage. Dan decided he would make the perfect calf roping horse.

Every day after evening chores, Dan would practice his calf roping. He and his new good horse had gotten adequately quick and accurate. Dan was ready for his pro-rodeo uncles to come home.

When the uncles were getting ready to head to the rodeo, they invited Dan to tag along. He'd been waiting for that very moment.

He appeared that morning in a sparkling white T-shirt, some well-experienced Wranglers and his work boots. Topping off this glory was a brand new straw hat he'd bought with his last paycheck. The hat was not your typical George Strait model but more along the lines of a Farmer Brown version.

The pro-rodeo uncles looked him over and said, "Good grief, boy, we're goin' to a roping – not to hoe cotton."

Dan smiled. He had a plan and thought he looked just right for it.

In the usual behind-the-chutes competition assessment, the calf ropers looked each other over and began estimating what kind of a fast time it would take to win the event judging by the look of those entered up.

Dan ambled by in his long-legged, double-jointed style with his new hat pulled down tight and leading his extra-large feather-legged horse. The serious ropers didn't even give him a passing glance.

Every roper, except Dan, had made their run, giving it their best shot. Those that did well, were already mentally spending the prize money. But there was one more cowboy to go.

Dan settled his big horse in the roping box, shot out as the barrier snapped back, threw a deadly loop, flew off his horse like a cat, flanked the calf and applied a lightning-quick wrap and hooey. Throwing his hands in the air, the clock stopped and the crowd went wild.

When the dust settled and the crowd quieted down, the announcer gave Dan's time. He not only won the event, he had set an arena record.

There wasn't a whole lot of talk about cotton farming on the way home that evening.

Marty Robbins made the "Cowboy in the Continental Suit" famous, but to this day when ropers gather for a cool one at the Chute One watering hole in Post, Texas, there is still talk about the kid in the Farmer Brown hat.

IT'S NOT EASY BEING FRUGALLY CHEAP

His pickup and trailer rig speak of a cheap – well, okay – frugal cowboy who makes do with what he has.

Last spring, Dan the team roper upgraded to a 1993 model.

Being the giving kind of guy he is, Dan donated his old truck, that he'd paid $600 for, to a friend who was in great need of it. Together they got it painted and overhauled and it was a gem.

Dan's new truck, the '93, came with chrome wheels. The friend with the old truck, in an effort to save Dan's reputation, offered to trade him the plain rims off the old truck for those chrome ones.

The horse trailer, a single-axle one-horse, is one you have to see to believe.

"Houston red" in color, as it was described to me, apparently denotes the rust because that's what it is front to back, top to bottom. It does have new reflector tape on it in lieu of lights.

Held together with baling wire, literally, you have to admire the horse brave enough to get in it.

When I first saw it, I thought it was a "just go down the road to the neighbors' to practice roping" trailer. I soon learned it was the main rig that travels to the major roping events.

I'm real surprised some of those ultra-fancy Texas arenas would let it in the parking lot.

I'd not be surprised if they asked him to park in the back, which wouldn't matter to Dan. His rig doesn't identify his roping ability.

Dan has his money allocated to specific categories – this much for Copenhagen, this much for beer, this much for horse feed. Whatever is left over is all his.

One time he had been saving for a pair of new boots. The pair he was wearing had been resoled a couple of times and were at their life's end, held together, more or less, with duct tape.

It had been a while since he'd shopped for boots and he soon realized his available cash fund, all $45 of it, wasn't going to cover the cost of new boots off the shelf at the Western store.

The weekly horse sale in Stephenville afforded him the opportunity to take a day off from his ranch job.

On the way, he stopped at an on-going flea market at the edge of town. Sure enough, there was a man with a stack of new boots in boxes.

Dan approached him and asked, "You have any size 13 double-E boots?" Both he and the man with boots were relieved to find one another.

Dan tried on the right boot and it was perfect. The man priced them at $30 and an elated Dan immediately began planning what to do with his leftover money.

He grabbed the boots, paid the man and left quickly before anyone could change their mind.

When he got to the horse sale, he decided to put his new boots on. The right one fit like a glove, but the left one, turned out to be a size 11. At this point, Dan had to decide if he wanted to go with one new boot and one old, or stay with both the old ones.

His pride and being tired of walking on gravel dictated that one old and one new would work just fine.

It was a couple of weeks before Dan could get back down to the flea market and could get settled up with the right size boots.

After he finally had both boots the same size, they lasted quite well and he's just now starting to think about duct tape.

"One more payment and it's mine."

HEELERS BY ANY OTHER NAME

Today we will look through the magnifying glass at team ropers, specifically the heeling end of the team.

There are three basic reasons that team-roping headers choose the heelers they do.

Top of the list is that they might actually have the skills to consistently catch the heels at which they throw their rope and, thereby, help win some money. This is the only reason ever actually stated in public, in front of people.

Not far down that list in second place is the fact that it helps if the heeler is fun and even better if he/she is a friend. It just makes being around them easy and entertaining, so why not enter up together?

This is actually the most common cause of most teams in most ropings today.

The third basis for selection, one that is not often discussed but holds true, is that a heeler might be selected for prestige.

If he has a famous, or even nearly famous name, a header might be inclined to dupe him into roping with him just to make the stories better later.

This last reason is illustrated by the sheer number of popular so-called "celebrity ropings" where, for an astronomical fee, a roper can be partnered with one of the "pro" boys from the Professional Rodeo Cowboy Association standings. This motive can be categorized as the "image enhancement" mode of roping.

The low-budget ropers have to dole out their funds evenly over the number of times they are allowed to enter a particular roping. Often a selected "hit list" is phoned far in advance of a roping to book teams to the best ability of the pocketbook.

When the funds won't allow for the "celebrity" status roping, the next best choice is image enhancement. A flashy-looking blonde on a prestigious blue roan horse will draw appreciative stares wherever she goes.

When you can round one of those up that can also catch, is fun and a friend, nobody seems to care who does or doesn't go to the pay window. "I had a lot of fun," is a good enough payoff.

Clearly roping is not about actually catching the cattle since, after all, they are already captured in a pen and then the chute.

And, they have been known to cause a certain amount of trouble getting them in and then out of the chute. Even that part of it can be downright dangerous as demonstrated by our resident heeling team roper Dan.

Short on chute-help on this particular day, Dan was going to run the remote control that opens the chute and lets the steer out.

It's about the size of a stop watch and hangs handily around one's neck on an unbreakable parachute cord. Handy, if you are just standing on the ground, potentially lethal if you are in hot pursuit of the steer in front of you.

As they left the box, the header neatly laid his loop over the horns of the steer, setting him up and turning left.

Dan handily laid his heel loop in front of the steer's fast-peddling hind feet and picked them up just like he knew what he was doing.

Headed to the saddle horn to dally as the slack pulled tight, he realized the tail of his rope was tangled in the parachute cord of the remote – the one that was still around his neck.

The header, not yet seeing the wreck about to happen, kept his horse in a powerful forward motion, taking the steer with him.

Dan, long-legged and lithe, bailed off his horse and began running behind the steer, rope in hand, in an effort to keep it from pulling the cord tight and hanging him in the arena dirt.

Fortunately, for all, the header saw "something" wasn't right as soon as he recognized Dan was afoot and somehow "connected" to the steer.

Everybody got pulled up, untangled and lived to tell about it.

The tales after the event were worse than the wreck itself. Dan, who is often labeled with undignified notoriety, won't live this one down for a long time.

Who says you need a tree to have a hangin'?

STRAW HATS AND NEW ROPES

Ranchers, ropers, and pretty girls are all getting ready for spring. Fruit trees are bursting in bloom and everybody I know is sneezing and sniffling from the annual pollen attack.

There are a few other signs of the impending season.

Dan has decided the winter wheat horses are about worn out and yearling cattle are about ready to ship.

His top pasture mounts, Slats and Thin Rind, have been used to doctor cattle, ride and work every day all winter.

It's time to take the cattle to town, get a paycheck, rest the winter horses and get down to the business of summer team roping.

Cue Ball is the professional rope horse on the outfit and is entirely too important to use in regular rotation with the wheat cattle.

Having spent the winter unemployed, he's fat and ready to win everything Dan can enter.

The practice arena has dried out, all potential headers and heelers have been invited to practice sessions and they have shown up in every state of readiness.

Some have new ropes, a few have new horses and the dedicated few spent the winter watching the champs on roping videos. It was their attempt to absorb some winning skills while running the remote control from the recliner.

Others simply outlasted the cold months by staying in shape with holding up their end of the cold mountain brewery distribution.

When Dan was looking for a new heel horse, Tex, the horse-trading uncle, told him to find a "blown-up" head horse and they would make a heel horse of him.

When Tex went to look at the horse Dan found, he agreed that Dan had done a fine job in the "blown-up" department.

It took Dan all winter riding the new horse the 8-miles round trip to and from work every day, tying him to a tractor all day (Dan works at an implement dealer) to get to him to quit rearing in the box.

Miles and wet saddle blankets are still quite dependable training methods.

Jeff had a little different worry than the other twine-twirlers. Last year he got caught in a bind, a little short of funds, and had to settle for less than the George Strait-special straw hat.

Everybody gave him a hard time all summer long for wearing this hat with a less-than-becoming crease that fit like a box stuck on his head.

Feeling like his buddies were being unjust with their taunts and jeers, he decided he wasn't going to let it happen again this year.

His friend Grady told him the local flea market had a new shipment of straw hats and were just what he needed at the right price.

Grady was impressively familiar with the local flea market.

Why, just last week he had bought a complete set of bowling pins at a wonderful bargain price. He had put them with the shower seat he had bought just the week before for a ticket price too good to be true.

Jeff recalled that last year his hat, the box-fit special, had blown off in the arena during a roping.

Two ropers, making a fast run at a steer, ran over the hat before Jeff could get it picked up. He also recalled that Grady had been one of them.

With this remembrance, Jeff became quite suspicious that Grady might have an ulterior motive to suggesting the flea market bargain straw.

He decided he would not let his Spring exuberance get him to just throw away his roping money, so he settled for a gimme-cap to start the season.

This is quite possibly the reason team ropers are traditionally not known for their cowboy fashion sense.

Let Spring begin!

DOING IT WRONG TO GET IT RIGHT

Life's lessons are rarely learned in the classroom and, for country kids, they often involve some sort of event that includes pain, disappointment and then, hopefully, an "Aha" moment for improvement in the next go-round.

My dad wasn't one to spend a lot of time explaining things. I didn't know any better as a kid, but as an adult looking back, I know I learned most of what I know by paying attention. By osmosis.

He wasn't loud or aggressive about letting us know if we, my brothers and I, did something wrong, but he also didn't often acknowledge to us if we did something right. We just had to figure it out by watching his reaction.

When he was pleased, he would tell my mother and she would, in the way mothers do to encourage, pass on the information that Dad was proud of us for something we had accomplished.

In this manner, I learned how to read a horse, watch a cow and know where to get to not be in the way when holding herd or working in the alley. He "showed" more than he taught about working with colts, riding with easy hands, a light handle, and when it was time to establish who was in charge.

Sometimes for kids, doing it the wrong way is the best way to learn how to do it right.

When Dan was young, about 7-8 years old, his grandpa lost the coin toss and was delegated the job of teaching him the art of calf roping.

They worked their way through the correct positioning with the horse behind the calf, loop size, proper swing and then the tie. At last, they were down to the part where Dan was to actually get on a horse and make a run in the arena.

Dan's excitement had been building, listening as patiently as he could while his grandpa explained each step of the process. With great enthusiasm, he got on his horse, backed him the roping box and nodded with his eye on the calf in the chute.

The calf flew out and headed rapidly toward the other end of the arena. Dan and his horse broke of the box, clean and smooth. He threw a big loop, catching the calf somewhere around his belly button.

Not quite textbook style, but a catch just the same.

He correctly made a quick dismount off the right side, only to land in a heap in the arena dirt.

Getting up, he tackled the calf, managed to string two instead of three of the four feet that were kicking around, and stood up with his hands in the air, triumphant.

About that time, the calf used one of his free legs to take a kicking swing, tripped Dan, and put him back in the dirt.

Undaunted, Dan jumped up and made a mad dash back to his horse, correctly not spooking him into a backward run. About the same time, the calf had managed to kick loose of the wrap and hooey and stood there shaking his head.

Dan got back in the saddle, rode up and took the rope off the calf.

When Dan got back up to the chute, he asked Grandpa "What do you think I should do next time?"

With a serious look and tone, grandpa said, "Well, I think you should take dead aim at that calf, get a good break from the box and run right over the top of him."

Dan stood there a minute thinking about that and finally asked, "Why?"

Grandpa told him he had done everything wrong that he possibly could on that last run and running over the calf was the only thing left.

Some days, it just seems that it won't get better until absolutely everything goes wrong first.

ALL IT TAKES IS A GOOD ROPIN' ROCK

Dan comes from a long line of rodeo ropers and he loves and trusts every one of them to tell him like it is.

Periodically, like every roper, he will get in a slump. The roping cattle take on a more relaxed appearance. They know they are in no danger.

When this happens, Dan heads to visit Uncle Tex for a quick fix to his problems.

Dan is on the close side of 40, and every time he talks to Tex he wonders why exactly Tex didn't tell him this particular magic tidbit when he was just a button. It's like his lips were sewed shut and Dan has to pry it out of him.

This educational meeting was set to take place at the local greasy spoon for breakfast. Tex showed up, listened attentively to the problem, said, "Hmmmmm." That was all Dan got.

After they ate and got as many refills on the coffee as the café allowed, they wandered next door to the local cowboy "toy" store. They were still visiting and inspecting all the newest inventions, like the Heel-O-Matic, Robosteer, Hot Heels, all the brightly colored ropes, new bits, new saddles and other essentials for the modern, techno-smart roper.

Dan was drooling, trying to figure which gadget would cure his current problem. Tex, of the old school, just looked.

They went on out to the ranch, still visiting and just fooling around. Like all cowboys, if they don't already got a fork in their hand, they pick up ropes and start throwing a few practice loops.

Not having gotten the fix to his roping problem, Dan decided to try again.

"You know, Tex, I'm kind of looking for a new horse."

Tex: "How come? Can't the one you got get to the other end of the arena?"

Dan: "Yeah, but I'm not roping too good on him."

Tex: "You ain't ropin' too good right now and there ain't a horse in sight. What's he do?"

Dan: "Well, he pretty regular bucks with me while I'm trying to rope."

Tex: "So ride him."

Dan: "He ain't terrible fast either."

Tex: "Lose your spurs, boy?"

Dan: "You think one of those new pink ropes would help?"

Here's where the well rope story came in. You all know that one where if you are any count as a roper you can rope with a well rope.

Dan was about to think Tex wouldn't be able to cure his current difficulty, when Tex unzipped his lip.

He set the low-tech sawhorse they were roping at an angle where there was only about an inch clearance to the ground. He put a rock just to the right of the heels and told Dan to get his elbow up, tip down, feel his rope with his little finger, knock the rock through and catch the heels.

Seemed simple enough. Dan has been home a week and hasn't missed a set of doubles yet.

Probably if old Walt would have had a real good rock like that, it wouldn't have taken him that extra year to win his eighth Heeling World Championship.

I'm headed out to find me a good roping rock.

DONNIE AND THE BUNKHOUSE

Some years ago, Dan signed on to work at the Tierra Verde Ranch. While Dan was moving his gear into the bunkhouse, Stan, another young hand, showed up and said they needed to go rescue a friend in trouble.

That would be Donnie, who was married and lived off the ranch.

Donnie was going to make a good old man because he'd been practicing it since he was a kid. Work and worry weren't going to be his cause of death at any age.

He was the kind of guy the boss gave the easy jobs to because, in his ineptness, Donnie wasn't about to complete a complicated project.

Dan had gas in his truck, so he and Stan jumped in it and off they went. They pulled up to Donnie's house and met his wife in less than desirable circumstances. She flew out of the house with an armful of shirts, threw them into the truck over the top of Dan and told him, "Don't bring that S.O.B. back here."

"Yes Ma'am," he told her.

She disappeared shortly, but when next he saw her, she was throwing a load of Wranglers over Stan on the passenger side and giving him similar instructions.

They were getting anxious about Donnie when he appeared running full tilt around the house in hot pursuit of a goat. The goat had a pretty good lead, but Donnie was hot on his trail. The third time around the house, Donnie made a flying tackle, broadsided the goat and got up yelling, "I'm the champeen goat dogger of all time."

He then spied his wife standing with her hands on her hips. At the same time, he saw the rescue truck so he cut across the flowerbed and dove into the pickup. Dan quickly put it in gear and they retreated.

Donnie stayed a fair spell in the bunkhouse with Dan and Stan. When he figured his wife had gotten lonesome enough and probably tired of cooking for one, he went back home and used some cowboy charm to reclaim his roost.

Dan was young and single and loved a good time. When payday rolled around, he would routinely travel to Stephenville to "give the girls a chance" at one of the local honky tonks.

One Friday evening, cold and misty, the boss told the boys to take it easy and go on home as they'd been working pretty hard. Donnie asked Dan where he was going. When Dan allowed that he thought the town girls could use some dance lessons, Donnie announced that he was going along.

Borrowing a clean pair of Wranglers and putting on his new sparkling white high-top Nike tennis shoes, he was good to go. The lads were welcomed at the local spirit emporium, as good-looking cowboys always are. They spent the night dancing and having fun but at closing time Dan told Donnie to load up and he'd take him home.

Down the road a ways, Donnie got very quiet and suddenly asked Dan to pull over. He bailed out of the truck and began rubbing green grass on his shoes until he had them covered in grass stains. Dan just had to ask, "Why?"

"I'm going to tell my wife the pure truth," Donnie replied. "I'm going to tell her we have been out dancing and chasing women all night.

She won't believe me. I can hear her calling me a liar and telling me I ain't been dancing, I been playing golf."

Dan thought about this awhile. Being a tactful person, he told Donnie that was a pretty good story but there was one flaw. There wasn't a golf course within a hundred miles and neither one of them had ever been real close to a golf club.

The bunkhouse got crowded again the next day.

THE TATER TOT EXPLOSION

Any cowboy will tell you that bachelorhood has its advantages, but cooking isn't always one of them.

A fella is usually pretty busy all summer – in a hurry and trying to get his work done so he can do his other stuff that involves horses, saddles, trailers and ropes.

The recent monsoon rains forced through the area by the landlocked hurricanes have left Dan the team roper fending for himself for days on end because there is no roping practice at his partner's and therefore no home cooked meals from his partner's wife.

For Dan, rain brings on some of the issues that become glaring in bachelorhood. No one to visit with except the dog, and while that's acceptable most of the time, there is also no one to cook for him except ... himself.

Proof of the danger in that came one night this week.

Leaving his work at the farm implement dealership quite hungry, Dan said he had stopped on the way home and bought a bag of frozen tater tots with a plan to make a tater tot casserole.

Upon arrival at his humble homestead, he placed them in the bottom of a casserole dish, added a can of Wolf Brand Chili on top and then a nice covering of grated cheese for the next layer.

Thinking his culinary creation was looking quite good, he added a few sliced-up wieners on top and then yet another layer of some diced jalapeños.

To his way of thinking, this had to be about the best supper ever.

Knowing he had piled a lot of food into the one dish, he shoved it in the microwave and cranked it up a ways, thinking it would take awhile to get it all warmed completely through.

He wasn't sure exactly how long he needed to set it for, so he allowed plenty of time for his masterpiece to get totally done.

Then, remembering he needed to go check on Pittsburgh's water, he headed out to the horse corrals while his delectable dinner cooked nuclear-style.

He got sidetracked, as cowboys are wont to do, and it was a good 30 minutes before he got back to the house.

What he found inside his kitchen was the aftermath of the complete explosion of his microwave and its contents.

There was chili, wieners and tater tots all over the ceiling with tendrils of cheese hanging in various places around the room.

His first move was to pick up the microwave and deliver it to the trash, knowing it would never be the same again.

Too tired to care much about the mess, his main concern was still the fact he was very hungry.

Like most cowboys in cow camp after long hard day, he resorted to the old standby – canned peaches.

He first drank off the liquid, then he filled the can up with whiskey, sat down, and ate his supper of "pickled" peaches.

None of this would have happened if it hadn't rained for days and days and if he had just been able to rope.

OPEN SEASON ON WINTER WARMTH

Since winter seems to be just around the corner, the subject of "meatier women" has frequently begun creeping into the cowboys' vocabulary.

It has to be a throwback gene to the cave man days and the survival of the guy with the "warmest" wife.

Dan the team roper is reminding his friends that he lives alone and has an old, drafty trailer house which is conducive to, and an incentive for, snuggling for winter survival. He is clear that, under those circumstances, the anorexic-type woman is completely out of the question.

With all his good friends, you are sure to know they will fix him up better than he could ever have imagined.

One idea was that he troll the buffet lines in town for the love of his life, but that scared him speechless when a photo preview was included with the email suggestion.

Another offered herself up with the caveat of considerable age, an extremely accurate cowboy B.S. detector and a large cast iron skillet.

Dan thought it might be OK, but she had to bring her own microwave. You ll recall he recently blew his up in the tater tot explosion.

Not long ago, Dan's ole buddy Donnie called him and said he was moving back to the area. Dan began recalling "the best of Donnie" stories.

Dan and Donnie were working on a ranch, but not living there. In money-saving mode, Donnie rode a Moped to and from his job.

A Great Dane on the route liked to chase the motor-scooter cowboy.

One morning, Donnie came limping into work, somewhat bloodied up. "You know, I have no idea if that Great Dane is still in the ditch underneath that Moped," he said.

At lunch, they took a ranch truck down to the crash site and sorted everything out. Dog 1, Moped 0. Donnie got a truck.

Donnie invited everybody over to eat one night. He was going to cook it up himself. The gang showed up and he handed them each a bowl teeming with unidentified ingredients. Someone finally asked. Ranch Style Beans and tuna fish.

When he called, Donnie told Dan he had been shopping in the horse trading magazines because as soon as he moved back, he wanted to take up roping again with Dan and his pals.

He said, as he understands horse ads, there are two categories, "He's a good'un," and "He has a world of potential."

"Don't want none of them potential ones, that sounds like work," Donnie said. When Dan asked him if he was looking for heading or heeling horse, Donnie said it didn't matter, he had decided to get the cheapest "he's a good'un" he could find.

Dan reminded him that he never was very good at roping and Donnie's response was, "Don't matter. I'm mostly in it for the beer drinking. I figure six or eight 12-packs and I'll have this team roping thing down."

When Dan asked Donnie how many times he'd been married, he said, "Four and half."

"How do you figure to have been married a half a time?" Dan asked.

"Well," Donnie explained, "I've been married four times in a church building. Then one time I got drunk in Oklahoma and had some sort of Indian ceremony with a fat woman. I might have been getting married then."

Likely, it was close to winter when that happened. Wonder if she had a microwave? Dan, perhaps, could give her a call.

THE BLAME GAME

Clearly, it is nearly spring. One indicator is the ropers' magazines and papers are teeming with new events offering incentives prizes of big cash, saddles, trailers and fancy pickups.

It is enough to send a roping addict to a padded room when the weather refuses to cooperate with his itch, and definite need, to practice. This adds immeasurable value to any warm, clear days that might show up after a long winter of miserable roping weather.

One such long, wet, cold spell was just ending and Jesse and Dan were planning on making up for lost roping time.

Dan and his prize roping horse Slats rode into the arena.

"Man, Jesse, the arena looks good."

Dan had not really looked the arena over, but he knew that Jesse always liked a compliment for the hour or so he'd spent running the harrow over the sand.

Even though that wasn't the case this day, Jesse took the compliment anyway and they proceeded to warm up their horses.

Because they were turned out to pasture during the cold weather, the steers were spunky and fresh.

Ready for the first run, Jesse backed into the heading box, pulled his hat down tighter even though his ears were already at a 90-degree angle, pushed his shades up, scratched his backside, tucked his rope and nodded for the steer.

Dan, on the heeling side, made similar preparations and, thinking to encourage his horse a little, told him "OK Slats, don't do nothin' stupid."

The first steer broke from the box as if he was on a too-tight bungee cord. Jesse reached and roped him.

Slats had not calculated the 'catastrophe response' necessary for this steer and was just a tick late. No problem, they had him captured anyway, faced and full of themselves with the success.

"Ain't nothin' to this roping – we just needed a little rest," commented one of the loopers.

The chute help, keeping quiet counsel, thought to herself, "Something comes to mind about fools rushing in."

Imminently, it would be apparent that the steers had all been "reprogrammed" during their weather-induced respite.

The next steer had developed a broken "rudder." He broke from the chute, took a hard left directly in front of the heading horse and escaped to the corner of the arena.

The one after that decided to come to a complete stop just as the head loop was launched.

The trend continued as a couple more steers showed that they, too, had developed head tricks that made life complicated for the ropers.

Jesse and Dan took a break to plot strategy, knowing there current plan was not working well. Their calculated decision was to let the cattle out of the chute a little further, trick them into thinking nobody was in hot pursuit and then snare them with a big seine-style loop.

With that design in mind, when the next steer was out, Jesse's head loop settled around the horns as soft as Mexico rain and he pulled the slack perfectly. They went left, made the corner and Dan scooped up the back feet with ease.

Cowboy style and grace will prevail every time. It will, also, exit just as quickly as it arrived.

Likely, bad weather roping days are not yet over, but ropers the world over are counting on spring bringing major-league success to each of them.

Meanwhile back at the roping pens, no doubt about it, the steers are plotting how to foil that grand idea. Ask any roper that had a bad run and, usually, his first blame will fall on the cattle.

"We'd have been a six if that steer wouldn't have ducked his head!"

THE DEVIL IS IN THE DETAILS

Dan and his roping partner Monte, are about as diabolically calculating as ropers can be – fine-tuning every facet of their sport and always looking for an edge that promises more frequent and lucrative trips to the pay window.

Last fall they decided they were a little weary of winning just enough at the local jackpots to almost get their fees back and were intent on setting their sights on any number of bigger, tougher ropings.

You know the ones. Thousands of ropers all vying to win a purple pickup truck with flames painted down the side. Prestige doesn't always partner with dignity, which, by the way, is usually hard to find at a team roping.

The duo practiced day and night and in all kinds of weather. When the early spring rains turned the arenas into rice paddies, they drove as far as they needed to find a covered practice pen.

No team roper has ever been accused of getting his priorities out of order as long as you agree his first priority is roping.

When the first "big one" rolled around, they entered up and arrived feeling skillfully prepared and confident. All things in order, when their names were called they backed their horses into the box with hats pulled down tight and minds focused on the task at hand.

Their first steer left the chute like a scalded dog headed for cool water. The cowboys traveled a good distance down the arena before they got him roped and turned, making their time a little long in the first round.

Their plan for the second steer involved Monte pushing the barrier a little harder and determining he would not play catch up this time. Of course, this steer broke from the chute and promptly came to a complete stop.

The judge culled that steer and gave the team another one who happened to be of the Nascar-bred Corriente line. They caught him but, again, their time wasn't anything to be too proud about.

The day progressed in much the same pattern. Another steer ducked under the heading horse's neck and made a beeline to the arena wall like he was on his way to the hot dog stand.

They caught him "long," but it qualified them for the short round. In the short round, with a last shot at making some money, their steer moseyed out of the box like an AARP member looking for a Furrs cafeteria.

The ropers pulled their horses up hard, flailed a couple loops at the gawking steer and in spite of the odds, managed to get a qualified time. It wasn't pretty but they had four steers down clean.

Their hard-earned, fine-tuned calculated efforts had put them pretty close to the bottom of the list of qualifiers. With no new money in their pockets, they loaded up and headed home.

The return trip, as it often does, offers enough miles and time for serious introspection.

It was during this analytical survey that the duo realized there was a critical element to roping they had neglected to factor into their program. Sure enough, they need to go home and practice their "luck of the draw."

YOU CAN'T MAKE THIS STUFF UP

Dan, of Slats and Dan fame, is an interesting character in his own right. He enjoys his happy hick persona to the limit and is all cowboy but with a real job to support his team roping lifestyle.

Working for the Ford-New Holland dealership, Dan is something of an expert at hay baling and tractors, has been through company schools and can recite parts numbers from a seemingly photographic memory. He supplements his income by day-working at different outfits and caring for wheat pasture cattle in the winter.

Dan tells non-stop goofy stories about his hillbilly relatives. He keeps his valuables in a little kid's tin lunch box in the floorboard of his truck saying they might get stolen if he leaves them in the house. This includes his "paw's" antique .22 saddlebag gun, Slat's papers, the house papers, truck title and his high school ring.

Dan's best friend since childhood is Tim. They split the rent for a couple years until Tim discovered all the benefits that came with them and selected a wife.

One morning before the wife came into the picture, they woke up and found their hats fit a little "hangover" tight. After some serious discussion they decided that maybe they should eat more vegetables other than hops, corn and tobacco.

They didn't want to go plumb off the deep end and cook anything – the gut grenades and belly-washer drinks at the local Quick Stop were filling most of the requirements for gourmet meals. After considering all their options in the health food line they decided on a cantaloupe.

Since they could get that at the stand by the side of the road next to the roping arena, everything worked out well. When they got home with their purchase they decided it might be a little green so they put it in a bowl and put it in the icebox. That was in 2002.

The next time they thought about health food was in 2003. The cantaloupe was still in the icebox but they donated it to the coon who hunted over the pasture fence. They cleaned out the icebox with a little kerosene and decided maybe health food was overrated.

Wallace is another of Dan's friends. A few years back, Dan and Wallace were at a mega-big practice roping and as both were very new to the sport, they had no clue what was going on and how things worked. They'd made a pact that if either saw the other doing something really stupid they would tell each other. That would avoid the embarrassment in front of a whole bunch of people.

It was Dan's turn first. "Wallace, you remember what we said about if we saw the other doing something really dumb we would say so?" Wallace nodded his recollection.

"I see you got some of those really high-dollar skid boots endorsed by Coopers on your horse tonight," Dan said. Wallace nodded. "You know those belong on the back feet of your horse?" Dan asked.

Wallace's logic was simple. "I know they are supposed to go on the back feet, but this damn horse won't let me put them on his back feet and they cost me $55. Any time I blow $55 on something for a horse, he's going to wear it somewhere. If he hadn't let me put them on his front feet he was going to have to wear them on his ears."

Wallace roped all evening in that manner, giving stern looks to anyone that may have glanced his way in question. No one said a thing.

Cowboy banter is basic, honest and always entertaining. Conversations run fast and in many directions – like the one about Kathy the Bling Bling Queen from Right Smack Out of the Crack, Oklahoma.

You just can't make this stuff up. It just wouldn't be believable.

COWBOY SLAPSTICK IS ALIVE AND WELL

Always a helpful bunch, the friends of cowboys will go to great lengths to "be there for you," especially if it provides them entertainment in the process.

Dan the Team Roper's recent situation made public, the one where he was looking for a fuller-figured gal to keep him warm this winter at his drafty trailer house, spawned, at his expense, a landslide of "fun" in Comanche, Texas.

"Oh my Lord," he said, "The hunt just got started to find me a meaty woman. The search seems to involve many."

That very morning he walked into his job at the farm implement and feed store and found, taped to the coffee pot, one of those photos that circulated the Internet.

It involved a considerably oversized gal who was unable to locate her little Chihuahua dog, that could only be seen if you were standing behind her.

Dan played dumb and acted like he didn't "get it," knowing full well the boys had read the story that very morning.

As the farmers began arriving to drink coffee and harass Dan, they began helping out in a number of ways.

One even offered his "fat wife for the winter." Dan kindly declined.

An email from roping buddy Blonde Sally offered to guide Dan in "broadening his horizons" with this woman search.

She had set him up an appointment for a Brazilian bikini wax. Dan was not quite sure what all that involved, but he was absolutely certain he didn't want any part of it.

Prior to this day, it had been raining for weeks and feed sales had been bottoming out.

Folks, in their eternal rural optimism, knew they didn't need and weren't going to need feed because a growth of new grass was most certainly on the way.

However, before 9:30 that morning, 64 bags of feed had been sold as it gave folks a reason to be at the store.

Stories beget stories. And so, Dan began telling a few more.

He launched into the tale of his honeymoon with his first ex-wife.

He had taken her to a farm auction in the Panhandle so he could buy a grain drill.

He regaled the coffee pot bunch with details of the adventure.

He told them that his blue heeler dog had hidden under the toolbox in the bed of the pickup and ended up going with them on the honeymoon.

Then the pooch proceeded to bite some guy at the motel in Plainview.

To top off the eventful trip, Dan said he ran over a parked Volkswagon in the motel lot in Lubbock.

"It was a very long two days and a large combo supper at Whataburger did nothing to soothe the situation. But I did get the grain drill home safe," he said.

Dan recalled that he didn't even really know she was mad about it all until seven months later when she made him go to counseling with her.

There he found out she had wanted to go to Cancun.

"What? And miss all this fun?" he asked.

The marriage didn't last long after that.

This story came on the heels of another roper celebrating his 30-something wedding anniversary by buying his bride eight new Corriente roping steers.

"Well, I guess it is his anniversary, too," she chuckled.

In the meantime, Dan is extremely happy.

Some guy wandered into the store and asked for his recipe for Tater Tot Casserole.

The recent culinary notoriety has moved Dan to take his cooking quite seriously.

Two days ago, he upgraded his kitchenwares with a crock-pot. Proudly he announced that it even came with a lid.

KNOW WHEN TO HOLD 'EM,
KNOW WHEN TO RUN

Kenny Rogers was wise with years when he sang, "You gotta know when to hold 'em," a song that finished with "know when to run."

Every cowboy with any amount of experience has been in that place in time when he knows it is just better to bail out, or off, to save his life.

Johnny had a new job and a new wife. His first assignment on the new ranch was to gather a fence-jumping bull out of a set of heifers where he didn't belong.

When he asked the cow boss which horse would be good enough to take on this mission, he was directed to a big, rawboned, feather-legged dun.

Johnny saddled the big horse, loaded him and his new wife, and off they went to the pasture. So far, so good.

Instructions for the new wife included watching him rope the wandering Romeo, after which she was then to bounce across the cactus, sagebrush and rocks with the pickup and trailer so Johnny could load the bull.

Johnny cut the bull out of the herd of heifers, roped him handily and waved to his wife.

Simultaneously, the bull decided to get on the fight. He made a run at Johnny and the dun and, in the process, he somehow got the rope caught under the rubber wrap on the saddle horn.

The horse remembered what the cow boss had forgotten to tell Johnny – that he would buck at the first opportunity he sensed the cowboy's attention was not fully on riding.

Johnny couldn't get his dally off, couldn't get the horse to quit bucking, couldn't find a soft spot to land, or at least, one without cactus. And worse, he couldn't hurry his wife up. At any rate, he was in big trouble.

Finally, he decided that it might be a good time to let the horse and the bull have it and he bailed off. The horse stopped bucking, the rope came off the horn and the bull came on the fight. "Know when to run."

The only thing that saved Johnny's life was that he was jumping over the sagebrush and the bull was going around them.

Eventually, the wife pulled the truck and trailer in between Johnny and the bull and he bailed in, but not without words of gratitude. Although a few years later, in divorce court, the wife mentioned she wished she had let that bull run over his sorry hide.

Dan had a big outlaw horse in his string named Cobra, named such because no tie-down in the world would keep his head from that "cobra" position. Dan had nightmares about Cobra coming out of a basket, swaying that head at him.

About that time, Robert Redford arrived on the big screen with the *Horse Whisperer* and Dan decided that laying Cobra down and sitting on his head might be the treatment of choice.

Redford looked pretty good doing that, he thought, and little children could ride the horse afterward, so it must work. Tim came to help, and beer was involved.

They roped Cobra, saddled him, laid him over and Dan was sitting on his side. Cobra objected and was fighting the weight planted on his side. The big horse somehow caught a hole in Dan's britches on the saddle horn and he could find no good way to get loose.

Dan was aware of the theory that you should always wear clean drawers in case you're in a wreck, but he didn't have on any drawers at all that day. He would have been happy to let old Cobra up but he really didn't want to be shucked out of his britches right then.

That's where the "know when to hold 'em" wisdom came in handy.

ARISTOCRATS OF THE RANGE

Cowboys are the aristocrats of the agricultural wage earners. As such, there are two or three things that are absolutely essential for their well-being: well-trained, prestigious-looking horses, a good rope and a nice hat.

Dan had the horse department covered with Slats. And he was following Tex's advice about roping.

Since the last problem he had was roping his heel horse's front feet occasionally, Dan went to see what Tex had to say.

Tex advised getting a rope his horse could see and then he wouldn't step in the loop. So Dan went down to the feed store, looked over all the brightly colored twines and came out with one of the new Roper Vision ropes – a glow-in-the-dark lime green.

This iridescent wonder came with a pair of matching sunglasses, the theory being that the roper would better focus on his rope when wearing them.

Dan didn't follow this logic as he wasn't the one having trouble seeing a regular colored rope, but most definitely, Slats wasn't going to get the sunglasses.

He decided that a little pasture roping might help Slats out with his difficulty of stepping into a heel loop. Tim came to help him out. The project would not be a wasted effort. They needed to doctor a few yearlings who had picked up something at the sale barn.

In the pasture, they rode through all the new cattle slowly and doctored each one who needed a shot. With the job well-tended to, they decided to practice a little to break in Dan's new rope.

The owner of the cattle had driven up to the pasture fence and was watching the entire operation; the cowboys were unaware of their audience.

After watching the flurry of activity through the initial doctoring, the cow boss walked out to the ropers and asked, "What you boys doing now?"

"Why, we're roping and doctoring the sick cattle."

The owner allowed he had seen quite a bit of roping going on but in the most recent time-frame, he hadn't seen any shots given.

Dan, quick with his wit, answered sincerely, "Last night I soaked my rope in a bucket of penicillin and there's no need to give a shot. And, besides that, we been looking at these cattle and thought we might need to check and see if some of them had a bone in their leg."

Good help is hard to find, so, begrudgingly, the cow boss decided to tend to some of his other business.

While he was putting his paycheck in circulation with the new green rope purchase, Dan also selected a brand new straw hat. He really hated to part with the extra money, but lately he'd noticed there were some quite pretty girls at the team ropings.

He also knew that one of the secrets to good looks is a good hat. Sometimes a new hat needs some breaking-in, so he was wearing it a little to get the sweatband broke-in before next Saturday's team roping.

This morning's job was to do some pasture shredding with the tractor in the horse trap. Dan got the shredder hooked up, was moving right along with the blades cutting down the old grass neatly when a little puff of wind hit and lifted that new straw right off his head.

The next thing he heard was "brrrrrrtt" as the new straw went through the shredder. A few choice words later, Dan shrugged and thought to himself, "Those pretty girls are just going to have to be impressed with Slats and the new lime green rope."

Two out of three is not bad for an aristocrat.

HORSE TRADING

THE ART OF HORSE-TRADING

The cowboy stood at the door in what, at one time in some other lifetime, might have been a red canvas jacket.

Today it was more on the aged side of pink, fringed with raveled seams and had a pocket that was holding on only by a creative thought. He was headed to town.

After some verbal exchange with his wife who knew he owned a better coat and reminded him of that, he drove off.

His rationale was that he was on his way to bargain for something and it was politically incorrect to go horse-trading, a term used for any kind of bartering in the West, in new or even salvageable clothes. The nature of the work to be done drives the wardrobe selection.

A guy I know has a ranch on the east side of the state that is free and clear of mortgages. He has enough money in the bank to stock it without the banker's permission. Always, at shipping time, this man would appear at breakfast dressed like the rest of the cowboys and stay that way until the cattle were penned.

Somewhere just after the gate closed on the last steer and the cow buyer who had been waiting at the road with the cattle trucks showed up, this rancher would transform into a pitiful creature.

He would be clad in a sweater with holes at the elbows, cuffs hanging by a thread, a hat that had been through the big war and boots with more holes than a prairie dog patch.

The cow buyer would arrive all clean and shiny with creases in his jeans like George Strait, driving a new Lincoln, a white, new out-of-the-box Stetson and wearing a big pinkie ring. The contrast was always appreciable.

Even though the actual numbers for the trade had been signed for months before and were never questioned, it never failed to come up in a conversation that the next time the buyer would have to do a little better.

Trading, bartering and making a deal, whether buying or selling, keeps a cowboy's heart pumping strong the same way riding a good horse following fat cattle in belly-deep grass does.

There is an art to "horse trading" that is passed through generations. It is honed to a sharp edge, practiced constantly, even on the wife and kids, and often has nothing to do with a horse.

However, if it is about a horse, it very often will carry beyond the monetary value of the animal. It is not unusual for the trade conversation to go on for hours when that transaction involves something other than cash.

Showing up at the trader's place with a trailer hooked to the pickup puts a spring in the step of the guy with the merchandise for sale. It is common practice to leave the trailer parked a couple miles away at the truck stop and arrive only with any other tradable goods to start the transaction.

Before the discussion of cash floats into the conversation, there may be a considerable amount of time given to offering other livestock, colt breaking, horse trailers, day work, saddles, pasture leases or pick of the new litter of blue heeler pups as part of the "deal."

It is fun to listen to horse traders while standing out in a dusty lot watching one "trader" pit his wits against another. A true loss to this unheralded talent is the increase of horses being bought and sold over the Internet.

An entire art and language is being lost and the Western world will be a less colorful place for it. But more than ever, the "buyer beware" phrase needs to be pasted to the pickup windshield.

VOCABULARY OF A HORSE TRADER

There is an entire dictionary's worth of phrases, sayings and quotes you can pin to the horse trading business.

The best advice for the buyer is to carefully discern the words they hear and look for what they may actually signify. Hidden meaning is a trademark of a seasoned horse trader.

For example, when the trader tells you, "This horse will let you do all the thinking," it really means he is big, dumb and heavy-footed. If he says, "For this one you just need to start a little sooner or cut across," he means the horse has two speeds, slow and slower.

When the trader tells you "he'll watch a cow," he could mean that he'll actually have the instincts to keep his eye on the cattle and have some quick responsive action. However, it could also mean he'll stand in the gate and watch them go by.

And a buyer should always look beyond the obvious. "This horse doesn't let much get past him" usually doesn't mean he is alert and attentive. It more likely means the horse will booger at a shadow or a bird flying overhead at a thousand feet. Riding uneventfully through rolling tumbleweeds and blowing dust will never be an option.

The horse described as having "a nice little cowboy lope" is one that is so rough to ride he will loosen your teeth fillings at a trot and, if you can ever get him in a lope, he'll jar your hemorrhoids up to your tonsils.

This type of horse can be described as having the ability to give a woodpecker a headache. I know because I own one of these.

The age of a horse is often disputed, especially if the horse has no registration papers for proof of age or origin. The ability to "mouth a horse" and read their age by the stage and condition of their teeth is a real benefit to the buyer.

But the die-hard trader will always justify a smooth-mouthed old horse with the line, "He's been in a sandy pasture and his teeth may look a little older from that sand grinding at his teeth."

Buyers beware when you hear things like, "He doesn't buck very often." My suggestion would be that even if you don't mind an occasional bucker, if the trader can't tell you exactly when he does buck, keep shopping.

Other things to listen for are the brilliant statements like, "When his nose quits running and his eyes clear up he'll be just fine," or "I usually don't have to hobble him to saddle him but he just looks better when I do."

In a moment of trying to dump a real mess of a horse, they actually will say things that desperate, even to people who know better.

Horse traders come in all sizes, shapes and classes much like used car salesmen. Some you can't trust and others you shouldn't trust.

Having a horse for sale and being called a horse trader is much like being a writer and being labeled a journalist. It is just not all that flattering.

SLIPPING THEM BY AT THE SALE BARN

Two riders were sitting in the alley at the sale barn waiting their turn to ride their respective mounts through the sale ring. It was sale day.

She was passing the time visiting about seeing a nice stocking-legged sorrel that she rather liked and she thought maybe she would go make an offer on him before he hit the ring.

The cowboy thoughtfully but kindly offered his opinion of that idea.

"Your horse here for a reason?" he asked.

"Uh, huh," she said. "Sure is."

"Well mine is here for a reason," he said. "And so, I'm pretty sure that sorrel is here for a reason."

The logic was loud and clear. Selling a horse at the sale barn usually happens because you can't, in good conscience, sell him anywhere else.

Sale barn marketing of horses offers absolution for passing on the hidden sins of a particular horse to an unsuspecting buyer, holding the seller harmless and guilt free.

"We are not responsible for your ignorance" is the underlying caveat.

Beyond the obvious concerns of wondering what kind of horse you will have when his drugs wear off, a whole set of interesting things start happening once inside the sale ring.

"This fine animal is as gentle as a dead pig and safe enough for your wife and children," barks the auctioneer.

"Your wife spends more getting her hair fixed than this horse is going to cost you."

It is important to get a good look at the person riding the animal through the ring. A complete circus act of drama and theatrics will unfold to the knowing eye.

A diverse assortment of folks will parade through the ring using a variety of tactics to enhance the horse's attributes or detract from the obvious lack of them.

The trick rider will stand on the saddle, twirl ropes and crawl under the horse's belly to assure you his horse is the best at everything you might ever want to do on him.

The crybaby will sob, tears streaming the entire time he is selling the best friend he ever had.

Old pro-dad will put all three children on the back of the horse to assure buyers this is a genuine kid horse. Buyers should make sure those are really "his" kids.

The sequin queen will ride through glittering so brightly, bidders will keep raising a hand to shade their eyes, thus driving up the price of the very ugly horse.

A consignor will sometimes pay someone else to ride his horse through the ring. This rider-for-hire guy earns a living making horses that nobody else can ride look good.

He will lope into the ring; spin the horse three turns to the right and then three to the left.

Everyone whoops, hollers and forgets that the average buyer would fall off right after the loping part of the show.

Occasionally, you will find some truth in the sales pitch.

At a colt sale, the auctioneer announced, "These colts have never been ridden or ridden at. Take them home and mess them up to suit yourself."

I liked it when a prospective buyer asked the seller what kind of mouth the horse had. The seller said, "It doesn't matter, I'm going to do all of his talking for him."

Buying a "finished" horse at the sale barn usually means someone is finished with him. The big money question is, "Why?"

HORSES I SHOULDN'T HAVE BOUGHT

In this series about the refined art of horse-trading and those that participate, I would be remiss not to share a few confessions about those horses I shouldn't have bought.

I, meaning me personally or any one of several people I know that should all know better, have been dinged by the horse trader on more than one occasion.

I received a response to the series that very candidly portrayed what I've been saying.

"Last year I bought a horse at the Spring Horse Sale for $900 and sold it at the December sale for $500 and never looked back. I finally got a set of shoes on him before the sale. It took two of us and a lot of drugs. We even gave the horse some. He never learned to neck rein but I did get the buck out of him long enough to sell him. Talk about an outlaw. He sure was purty, though!"

I personally bought one of those really "purty" ones that was represented as a "little cinchy once in a while." It wasn't long before I realized I owned a horse that needed a shot of drugs before you could saddle him. He only flipped upside down when you pulled the cinch too tight or too fast and that only happened once in awhile. Sometimes he waited until you were sitting in the saddle.

Then there was the big, very pretty palomino that was the answer to a dream. It had rained a foot in the Panhandle, something that rarely happens. So when the cowgirl went to look at the horse, he was standing knee deep in mud. She fell in love with him at first sight, wrote the check and trudged him through the mud to the trailer to take him home. On dry ground, she could see he was about as pigeon-toed as he could be and still walk. Since it didn't rain again for a long time, it took her a while to find him a new home.

The other pigeon-toed horse story is much the same, except this big gentle giant was standing in knee deep grass in south-central Texas. He came with high recommendation for his gentle ways and since his job was to raise four ranch children, he made the trip back to New Mexico. The worst part for momma was trying to explain to her husband why she paid perfectly good money for a horse with front feet that were looking at each other.

Then there was the nearsighted barrel racing horse. He could and would turn like a rat in a barrel but the problem was he would do it about 10 feet in front of the barrel. Hard to win a barrel race that way.

And the mare that was bought at the racetrack with a head exactly like a mule. When the wife took the boss to see the horse, she had the owner back the mare out of the stall. The mare looked like a million bucks –all the way up to and until very long ears.

The sale-ring horse, that would kick your head off your shoulders if you surprised him, went back to the sale ring. Another friend said they have two horses they bought from a trader that will take a trader to get rid of them.

It seems to be human nature to fall for "purty" and ignore every warning signal that sets off alarms in us, telling us to move on, don't buy this one. I think humans have a tendency to pick their life mates the same way.

Someone said to me recently that it is easier to abandon a man than it is a horse. You don't have to worry about who is going to feed him.

The most honest reply I got when asking about the horse they shouldn't have bought was, "Really, almost every horse I ever bought I shouldn't have."

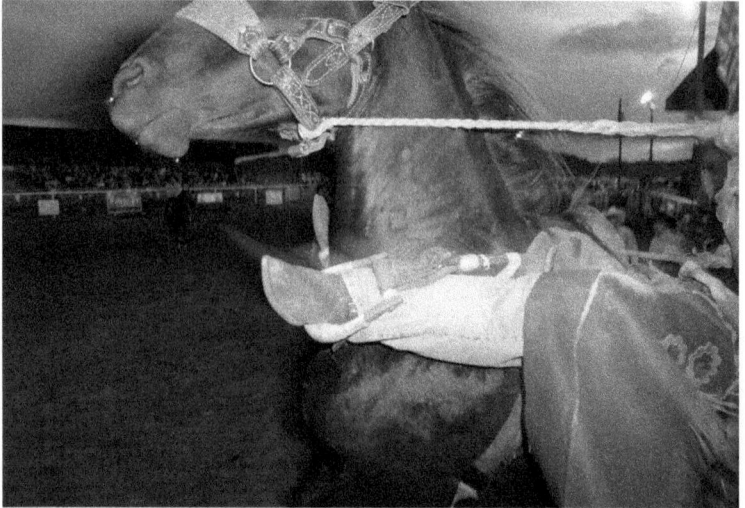

$27,000, OR BEST OFFER

Cowboys are born with a trading gene. Usually this involves swapping horses, livestock, trailers, saddles or pocketknives.

Horse trading requires a special language. If both parties in a trade are cowboys, the buyer is already in "beware" mode. However, for those who were not born down dirt roads, here is an example of a few choice phrases of trading vernacular used mostly in print advertisements.

"Very alert 12-year-old gelding, foundation stock, strong, heavy-muscled, will watch a cow. Friendly nature, quiet in the arena. Must see to appreciate. $27,000 or best offer."

The literal translation is:

> • Alert – he will spook if a bug within five miles moves. Nothing is going to sneak up on him.

> • Twelve years old is about the age where horses can no longer be positively aged by their teeth. He could be 34.

> • Foundation-bred means he looks exactly like a mustang and was adopted from the BLM in their effort to preserve the world before the wild horses eat it up.

• Strong means your best antique, foot-long, 40-pound Mexican bit won't hold him.

• Will watch a cow means he will watch the cow go right by.

• Friendly nature – he will pick your gloves out of your back pocket as well as gnaw on everything in the barn and everybody else's saddle if tied next to another horse.

• Quiet in the arena – he won't wake up before the cattle get through the stripping chute at the end of the run.

• Quiet in the box – he will sit there until next Friday if you don't liberally apply the spurs when you nod for your steer.

• Must see – the seller is hoping to get you to their pen, lock the gate and not let you out until you buy something.

• The price – that's always a starting place. Actually, the guy would be happy to see $800 and that horse's backside out his gate.

The trading world has three basic components: sellers, buyers and tire kickers. The variety of descriptive phrases applied to horses would enchant any clever wordsmith.

"Not the prettiest head you ever saw, but it's full of cow sense." That means his head looks like a pump jack, is exactly the same length as his back, and it would give most horses whiplash to hold it up.

"This is an after-dinner horse." A compliment, to mean that he is a smooth trotting horse who won't jiggle your dinner. It also likely means that he can trot all day in the shade of one tree.

"He has a smooth little cowboy lope that you'll love." This is supposed to infer that he can cover the miles smoothly. Nobody mentioned that it takes the first five miles to get him worked up to this cowboy lope and only 15 steps for him to fall back to that teeth-jarring trot.

"This is a horse who will let you do all the thinking." A good bit of this required thinking will also involve your spurs.

"All that chrome really dresses him up." This horse has enough white that the Quarter Horse Association referred the owner to the Paint Horse Association, and he has white hooves, which will require that he have shoes every day of his useful life.

A buyer might comment that the horse "looks a mite lively." To which the seller will reply, "He's just fun-loving. Don't take that little bucking streak too seriously."

When faced with tire kickers or reluctant buyers, the sellers will sometimes offer incentives.

- Horses are just like bananas. They get cheaper by the bunch.

- Better buy him today; I'm just like an old cow, get up in a different mood every day.

- "He's an easy keeper, and we've had him turned out for about six months resting him. He may be a little fresh." Which really means this fat, fresh horse hasn't been touched in the six months it took for the seller to get his leg cast off after the last time this horse bucked him off.

- "This horse can be ridden by little kids." This invariably refers to the little kid who was raised at the head of the creek and whose normal working attire includes a turkey feather in his hat and his britches tucked in his high-top boots.

When all the insults and high-flying rhetoric are over, generally both the buyer and seller will be secretly happy. Both will be heard down at the feed store in a day or two describing exactly how they got the better of the trade.

Trading is an art, and nobody does it better than the cowboys.

MATCHING THE SALES PITCH TO THE HORSE

Tex was a horse trader, skilled at many things, but horse-trading was his first love.

He is also Dan's uncle. You recall Dan of "Dan the team roper and Slats the politest horse ever" fame.

Living in an area where there are lots of cowboys, lots of ropers and lots of horsemen, Tex was making a viable living at trading horses. All those horsemen and ropers felt the need every once in awhile to change colors or upgrade their horses, so they'd go see Tex.

Tex tried hard to maintain an adequate inventory of trained horses but also made a point to fit each customer's needs to the abilities and personality of each horse. An admirable quality in a horse trader, indeed.

Calf roping was the "hot" sport at this particular time. Tex had Dan living in the bunkhouse, training and tuning up the calf roping horses for the customers.

Dan would get them going good, correct any little bad tendencies and be ready at a moment's notice to show these good horses to prospective buyers.

One afternoon, Rocky, a regular customer, called and told Tex he needed a new, but very good, calf horse.

Tex was caught kind of short in his "good calf horse" inventory. All he had was one sorrel stocking-legged horse standing in the lot ready to sell.

Tex told Rocky this good sorrel horse was quiet, had nice conformation, an impressive set of papers validating good breeding and had been started right. He would need some more polishing up on his arena work but would be a good horse someday. Tex priced the sorrel at $1,250.

Rocky told Tex he was on a winning streak, had won a couple of buckles lately and better yet, his girlfriend was real impressed with his roping. He really thought he needed a horse that was quite a lot better than the sorrel Tex had offered him.

Tex thought about it a minute and with the presence of mind only the best of horse traders can muster, he began describing a stocking-legged horse to Rocky. This one had lots of "chrome," the best conformation with lots of muscle definition. He guaranteed a good solid stop on the horse where a man could get off on the right, run down the rope and tie a very quick calf.

Tex promised this horse to be a surefire winner in anybody's book, but added the caution he really wasn't planning on selling him because he didn't have his registration papers. Reluctantly he priced this special horse at $2,500.

Rocky said he'd be right over, that this horse sounded like just what he needed. Dan showed the horse to him, demonstrating his skills on several calves in the arena, catching them all.

Rocky was in love. He'd never miss another calf, would have to build a trophy case for all the buckles he would win and his girlfriend would be more than impressed.

As Rocky drove off with this exceptional, you guessed it, sorrel stocking-legged horse in the trailer, Tex looked at Dan and said, "Well nephew, we are plumb out of horses now and while I hate to tear up these papers, Rocky ain't going to need them anyhow."

PANHANDLE COWBOYS

PUNCHING COWS IN THE PANHANDLE

Cowboys have their own style, and it is one that has evolved through a century of working their trade. Though widely imitated by everyone from John Travolta to a Wall Street wanna-be, the true essence of genuine is never quite captured.

Often the differences between imitation and genuine are so subtle only another cowboy will pick up on them. Cowboys often judge the other by the first impression. First it will be by the hat and boots on the cowboy and then the tack (saddle, etc.) that the horse is wearing. After that, the real test comes when observing a man's skill with a horse or his handling of cattle.

Unique to the occupation, cowboy style will vary every hundred miles of geography depending on weather, terrain, types of cattle work and necessities of the occupation.

The Texas Panhandle is said to draw the largest concentration of cattle on feed anywhere in the world. That makes it about the best place in the world to catch the largest number of working cowboys in the same place at any one time.

Some come from ranch-owning families and some from working-ranch-hand families. Some come from South Texas, some are buckaroos from Nevada, some have waded the Rio Grande, having cowboyed all the way from Chihuahua to El Paso.

Some come from places where towns crowded them out and some are kids working their way through college. Some have diplomas and are paying their dues at the bottom of the ladder before they go on to manage one of the mammoth feed yards.

Occasionally, one will have come West from the piney woods of southeastern United States after deciding he was tired of driving cattle trucks hauling stocker cattle to the Panhandle. In the winters they will drift down from Montana, The Dakotas, Kansas and anywhere the climate is not so forgiving.

The major portion of the Panhandle cowboys are homegrown Panhandle ranch hands coming to "town" to work the feedlots for awhile. They can always count on steady work, a steady paycheck and almost none of them have any illusions about the romance of cowboying.

The best of these work-hardened cowboys will be accustomed to recognizing sick cattle, handling all cattle gently, counting accurately, riding through the long days without much complaint. All will know how to process and doctor cattle. And, while most will hate that part of the job, they will be more than conscientious in their care of the cattle in their charge.

This melting pot of cowboy types, unless raised at the feedlots, will be overwhelmed with the sheer numbers of confined cattle. Most will have come from places where sections of land scattered cattle far and wide. They will bring their good horses that will likewise be appalled by the dust, mud and endless multitude of gates to be opened horseback.

The Panhandle feedlot cowboys are a colorful lot and perform an absolutely critical function for the cattle industry. Owners of the cattle in the feed yards, feedlot managers and feedlot owners recognize that these men are the backbone of this labor-intensive operation.

The cowboys themselves just look at it as their jobs – another part of being a cowboy. In good cowboy style, they just enjoy being punchers.

Next, I will detail the varieties in dress, tack and attitudes that come with cowboys from different parts of the country. In spite of their geographical differences, they all have one thing in common when they work in the Panhandle. They are the punchiest bunch of punchers you will ever see.

I SEE BY YOUR OUTFIT

There is a phrase made popular in song that says "Don't call him a cowboy until you've seen him ride." It goes right along with the wisdom of "clothes don't make the cowboy."

With the growing popularity of "cowboy" symposiums, cowboy poetry gatherings and other such galas made popular by permission of the urban cowboy craze of the '70s, the world has seen some amazing variations in what a cowboy is supposed to look like.

Let me first say, most of what the "world" sees on their side of the cattle guard "ain't it."

There is a melting pot of cowboys formed by the migration to the cowboy work available in the Texas Panhandle.

When the ranch-raised seasoned cowboy arrives in the Panhandle, his clothing and tack show a regional influence of where he calls home.

The South Texas cowboys, accustomed to dodging through thickets where everything has dangerously sized stickers, usually have tapaderos on their saddles, long leggings, lots of rawhide tack and a hat that will pull down real tight.

Their heavy-made stout horses will be startled by open country, gentle fat cattle, and they will spook at their own shadow when coming out in the daylight. These brush poppers will be surprised how exposed to the elements they seem to be after working in country covered solid in thorn trees and cactus. And, of course, the Texas Panhandle is infamous for its "elements."

The South Texas cowboy will have a saddle with a high cantle, complete with scratches for a signature of its life in the brush. Often they will have custom tack and silver on bits and spurs reflecting the pride attached to cowboying in that rough country. Every one of these brush hounds will be wearing a brush jacket whether it is a snowing blizzard or 112 degrees in the shade.

Nevada buckaroos will express an initial opinion that people in Texas or almost anywhere except Nevada do entirely too much work on foot. Buckaroos are generally too important to ever get off their horses and just don't see the sense in doing anything that can't be done on horseback. They mellow out after a while but, in reality, are surprisingly good at doing some unusual activities from a horse.

Their style with tack, saddles and clothing will reflect vanity as well as functionality. Buckaroo saddles may be the A-fork style, often with bucking rolls or a Wade tree style (sits low and stays put no matter what) with flat bottom stirrups with a strip of leather sewn in the back tread of the foot to help from losing a stirrup.

Their California mission-style bits will often be Garcia or Sliester made with slobber chains. Some will be wearing suspenders and most will have wild rags (large silk neck scarves) and low crown hats with flat brims. Most will be wearing 16-inch top boots with an under-slung heel. Their britches from the knee down will be several shades darker, never having seen the outside of their boot tops.

The vaqueros from the blue mountains of Mexico will come in without saddles or a horse, with well-worn clothing and not much else. They will be good with spurs, riatas and señoritas. Feedlot managers are often hesitant to hire these men but, when a good one comes along, he will be one of the best with horses.

Many times, though, these hard-working people are relegated to cleanup and processing crews. They have a history of having a grandmother in failing health who requires a visit about every three months and a return date is never more than a "maybe-so."

Just as often, entire families will return faithfully year after year to a good manager. They are as important to the industry as the very best college-educated managers.

Next, I will outline the defining rigs, garb and attitudes of a few more of the "boys in boots" from around the country who end up picking up their mail in the panhandle of Texas.

COWBOYS –
PACKAGED PRIDE IN WHAT THEY DO

Extreme weather is, and always has been, a noted feature offered up by the Texas Panhandle.

There are more stories about the "extremes" of hot and cold in that area than there are cattle in the feedlots. It is not uncommon to have someone tell you about their kin that moved to Amarillo, Texas, from Fairbanks, Alaska, and promptly moved back to Fairbanks after declaring it was "too damn cold" in Amarillo.

Blizzards, both sand and snow, are a common occurrence, and you just about have to be from there to like it there. But the close proximity to feed and cattle keep the feed yards in business and the cowboys employed.

Men from the smaller backyard cattle-growing region of the Southeast will come in with a woods colt, an old saddle and pointed-toe boots common to that region. With an excited attitude, they perceive this feedlot cowboy job to be an adventure.

The climate and severe working conditions of the feed yards can disillusion them in a hurry. They are able to go back home to a job and often do; resuming previously thought-of mundane careers of driving lumber or cattle trucks. Upon their return, they will be able to talk about their time working in the big country and how cold and hard it was.

The homegrown Panhandle ranch cowboys will have handmade tall-top boots with several rows of stitching, under slung heels and round toes to prevent stirrup hang ups.

They will bring handmade Western working saddles which are traditional-style double rigged and have a breast collar. Their saddles will wear anywhere from one to a dozen piggin' strings (small ropes) hanging from them as well as hobbles and medicine bags.

Their custom spurs will have hinged buttons and their spur and headstall buckles will be rust-colored with silver brands or initials. The spurs will sometimes bear cutting, roping or other type of special rowels reflecting the cowboy's particular interest. Some will sport the pizza-cutter rowels and jingle bobs that are usually to make noise to aggravate everybody else.

Panhandle punchers will, almost to a man, always wear black well-worn hats. These make a statement as well as any words in the English language verbalized could by saying, "I'm here now, move it on over." For the most part, that is not brag – just fact.

Arizona, New Mexico and Colorado cowboys all have variations in their style depending on the regions. The northern boys will have leggings for chaps instead of the short chinks worn by many from the hotter desert regions. New Mexico seems to favor the higher-topped boots and other garb that lets them blend across the borders in any direction.

Then there are a few idiosyncrasies in cowboy gear that each generation hopes will pass without becoming history. One of those currently is the "taco hat," which is a straw cowboy hat with an extremely wide brim folded up to resemble a taco shell. The style appeared to originate in the Panhandle but has begun migrating West. Recently, I heard it called the 'No Mirror" hat, meaning if that cowboy had a mirror and saw how ridiculous he looked, he would take it off and stomp on it.

While the Texas Panhandle cowboys will lay claim to be the punchiest bunch of punchers ever, the title will always be challenged by cowboys from everywhere.

If a cowboy isn't anything else, he is a walking, talking package of pride in what he does – no matter where he does it.

RANCH WIFE WOES

RANCH WIFE 201 –
THE HUNDRED-YEAR DIALOGUE

It is simple and it's predictable. The following is an upper-classman study for the ranch wife that has graduated from Ranch Wife 101, and is in the golden years of her tenure at the ranch.

When he says:

- "There is nothing in this house to eat."

That suggests that the aluminum foil covering the very many dishes in the refrigerator left from yesterday's branding are making them invisible. The same foil will not have that effect if covering a pie or cake on the counter.

- "I'll get that heavy box for you – just give me a minute."

This comes with the same flexible time span as his, "We'll be right back."

- "We're going to have a good calf crop this year. You should go with me and look at them, they're so cute now."

Translated, that reads: Bring those little hands of yours, we likely have some calves to pull after last night's 4-foot snow.

- "You're really getting to be a good hand horseback.

You deserve a better horse now." Dead giveaway for his plan to buy a new horse that he's eyed and then promote you to something he's already worn out.

• "You can always see what is going wrong with my roping. Would you come out to the arena with me and help me figure it out?"
A master at subterfuge, he is saying the neighbor that he was counting on to run the chute has canceled out.

• "Since you're going to town anyway to get supplies for the cattle working, would you pick something up for me?"
Certifiably, this list will require possibly two pages of a Big Chief tablet and will include something large and covered with grease, going by the bank to sign a heart-stopping bank note, vaccines that must be kept cold and beer with the same requirement, a couple of new ropes of a particular lay which every store in town will not have, and a widget, for which he has forgotten the correct name, from the NAPA store.

• "How much do we have in the checking account?"
This is not actually an inquiry of the current financial status, but has two possibilities of translation.
1. He's found something he cannot exist without, has already written a check for it and is looking for a good way to let you know the account is likely already overdrawn.
2. You are backed in the roping box, focused on some earned R & R, but he is tired of heeling for you and is wrangling a way to go do something he'd rather do.

• "What did you do with my ..." Fill in the blank here. This could be anything from the D-9 cat used to push brush to a small gizmo fix-it for the roping chute.

In reality, it indicates he has misplaced something and would like help finding it. In any event, absolutely the only answer you can give is that you never saw whatever it is, in your entire life.

• "I'll eat some of that if it would make you feel better."

Sometimes his efforts made you feel so good, it required making another, whatever it was, to feed the company it was intended for in the first place and who are arriving within the hour.

• "Did you open (or shut) all those gates when you moved the cattle to the back pasture this morning?"

This usually comes right after you have gotten to bed at the end of a long, long day. The guarantee is that it will make you lay awake all night and question yourself.

It will also reveal that after the first 100 years of marriage, he still thinks you don't have enough sense to do things right and actually thinks you would admit to it if you didn't.

I have been told there is a measurable amount of dignity in silence. I am still working on that dignity thing.

THE PRE-APOLOGY PLAN

Like most of the critters at the ranch, some cowboys are smarter than others. Tom happened to be one of the smarter ones.

He and Sue Ann worked together almost every day on a big ranch. He had figured out that at some point during the day the cattle, the horses or the hired help would do something that would require him to yell at Sue Ann.

I've mentioned before that a cowboy has this tendency to bark orders or a correction at his wife so that the people that need to hear it won't be offended but will still become informed.

Sue Ann's historical reaction to this tactic wasn't something Tom recalled with pleasantries. In view of the fact that most days she was the best help he had, he considered several options of minimizing the effect of his methods.

One morning at breakfast he decided to "pre-apologize" for any mistakes he might make during the day.

A pre-apology could and would cover almost everything, save time during the working day, and enable him to be comfortable in his recliner at day's end while she fixed supper instead of using that time to soothe her at the saddle house.

What goes up must come down is not a solid physics theory when it comes to cattle, and especially yearlings.

Sometimes the mistake of parking a semi-truck of cattle to be unloaded pointed downhill makes for an exciting moment. Nothing short of a thunderbolt will stop the stampede before they get to the bottom.

Give those same cattle a hill or mountain to climb and they'll find the top and take up permanent residence without any intention of coming back down unless, first, it becomes their idea.

When it's their idea, especially if you get a bunch of them gathered up on the top of the hill and pointed downward, refer to the previous paragraph and thunderbolt theory.

On this particular day, fall was approaching. Tom and Sue Ann had started moving the cattle from the high country to shipping pastures at lower elevations. The hired help had gathered up as day was breaking, slickers tied to the backs of their saddles for the inevitable afternoon showers.

Tom, Sue Ann and crew worked their way up the mountain trail to get above the cattle.

Yearlings can be pretty snakey in the brush. They have been known to sneak around behind the riders or even end up on the side of a cliff-like place where the only method to get them to move is the very un-cowboy method of throwing rocks at them.

That is one of those cowboy skills you don't hear talked about much.

When a sizeable bunch had been collected to a clearing, the hands started moving the cattle down a steep ridge top complete with plenty of rocks and deadfall. Sue Ann was riding point to the left of the front of the herd, charged with keeping them headed in the right direction.

As will happen, something that nobody but the cattle saw or heard spooked them, and the race was on. At a dead run downhill they raced but, of course, not in any direction they needed to be going. Sue Ann jumped up in the front of her saddle and rode hell-bent-for-leather through the treacherous steep terrain trying to head the cattle. In a effort to not kill herself or her horse in the rescue, she did have to pull up in a few places and select a less lethal path through a natural gauntlet of dangers.

Finally, a little further down the hill than Tom had planned, everything came to a halt. Tom, already calculating the pounds and dollars that just ran off the cattle, simply couldn't help himself even when he knew better. He rode over to Sue Ann and asked her why she let the cattle run and why she didn't get them stopped sooner.

It was good thinking on his part that he had already pre-apologized.

DID I FORGET TO MENTION THAT?

Men, in general, have a built-in gene making them masters at forgetting to mention important details that often dictate the outcome of a situation, up to and including the moment they could lose their lives or an important part of their anatomy.

The obvious incidents include forgetting to mention the existence of a wife, or some wild tale about why they didn't arrive home until the day after they were expected.

Cowboys, however, have different types of "Did I forget to mention that?" stories.

Ranch stories of this nature will sometimes involve a simple request to the wife along the lines of "Could you go get our black bull out of the neighbors' pasture today?"

What the head cowboy may have forgotten to mention is that once she finds the black bull in the four-section brush pasture, she will likely have to break up a fight between him and the neighbor's resident bull.

Then she will have to persuade the black bull that he would prefer to leave the neighbor's young heifers so she can drive him back to his ranch home. On the way out, she'll have to repair the fence that he tore up running away from home.

Being a sensible wife, she will know that the black bull, which is usually cooperative, will need to come to the pens at the headquarters to discourage the same scenario from happening all over again.

These kinds of projects are common to the status of "ranch wife" who is not usually surprised by the omission of finer details of the request.

Instead, a fair amount of get-even plotting will occupy the span of time it takes for the ride over to the neighbors, as well as the return trip.

When calves are shipped from the ranch, a permit from a state brand inspector is part of the process. The inspector in this case was about 5 feet tall and wore a pistol that came down almost to his knees. His demeanor indicated that he failed to recognize he was not God.

The ranch boss asked his wife to go help the brand inspector count and sort the calves, penning the heifers and steers separately.

The calves, at one end of a long corral alley, began to file by the little woman so she could determine their male or female status before directing their destination.

As the calves peeled away from the bunch, their speed of departure picked up, making the "viewing" considerably more difficult. The brand inspector suspected he would have to come to her rescue.

The cowgirl wasn't the least bit nervous about this assignment, and expeditiously called to the inspector, "In," for the heifers, and "By," for the steers.

A couple of hundred calves were sorted very quickly this way, with no slowing of the steady stream of cattle down the alley.

When it was all done, the inspector told the cowgirl he'd never seen anybody, male or female, sort cattle that quickly.

What he had failed notice was that all the heifer calves were specifically earmarked. To make her call, all she had to do was glance at their heads as they came toward her.

She figured it was information he didn't particularly need, so she "forgot" to mention it. Thereafter, she enjoyed a reputation as a very astute and competent cattle woman.

No mention was ever made that she shared the same "forgetfulness" indicative to the male species of her profession.

That quiet fame happens a lot at the ranch.

PUTTING THE DEFINITION ON NORMAL

Cowboys are generally reasonable critters, but sometimes their perception of reasonable and normal gets them in a bit of a bind.

Jerry and Donna, husband and wife, were driving down the road to pick up yet another load of wheat cattle to put out on pasture. This particular morning, Jerry was explaining in great detail to Donna just what his "reasonable expectations" included.

All he wanted was a normal wife, he said. A normal wife would not think crackers were an acceptable substitute so would not run out of bread. A normal wife wouldn't expect much of a poor wheat-pasture cowboy.

As is often the case in the Texas Panhandle, the weather was despicable, cold and raining right straight down. When they arrived at the pasture where their cattle to be picked up were penned, Jerry pulled up to the gate and waited. Nobody moved. Finally, he asked Donna, "Aren't you going to open the gate?"

Donna explained, "No normal wife would get out in this mess and get her boots muddy or her hair wet."

Jerry got out and opened the gate with no comment.

He backed up to the chute, got out and opened the trailer gate but as he headed back to the pickup, the wind blew the gate shut. This happened a time or two more until Jerry got a stick to prop it open.

Like a normal wife, Donna sat warm and dry in the cab of the pickup. As Jerry got back into the truck, the stick on the gate slipped and it banged shut. Finally, Jerry said, "OK, you made your point. Please get out and hold the gate for me."

Later, thinking to make up, Jerry told Donna that he had noticed she had a flat on the inside dually of her feed truck. He told her if she would take the tire off, he would fix the flat while she fixed lunch. Donna said she thought a normal wife would be taken to lunch.

That explosion passed as Jerry realized they still had all afternoon to go. They could put out another set of cattle and get them settled before dark. His thought was that he could go pick them up, Donna could go build a half mile of hot-wire fence he hadn't had time to get to and then they would meet to put the cattle out.

Donna said she had other plans for the afternoon. She outlined a normal-wife strategy of going to the mall and perhaps getting her hair and nails done. Jerry could then meet her and take to the new restaurant she had heard about.

The following day there were more cattle to get to wheat pasture. When this was done, Jerry noticed the hay trailer needed loaded as snow was forecast. Donna had always tended to this project.

Jerry wasn't that slow. He had become accustomed to her doing certain things. Today, so far, had contained no reference to "normal wives" and he had forgotten what had prompted him to mention it the day before. Donna informed him that all four tires on the hay trailer were flat and she recommended that he do something about it because flat tires didn't fall within the parameters of normal wife expertise.

It didn't happen often, but he gave up. He told her he could live without a normal wife, just please get back to doing what she had been doing. And, while she was at it, when she took the flats off and had them fixed, she may as well rotate the tires around the trailer too.

As a normal cowboy's wife, she got right on it.

THAT'S WOMAN'S WORK

Chauvinist is a cowboy word. You won't hear them say it, and most likely, without a little help, they can't spell it. However, they live it with a subtlety that defies description.

In the heart and mind of a cowboy, there is a long list of things that fall under the category of "woman's work" and even if they have to be sneaky about it, they are determined to make it her job, forever.

One of the most common frailties he will portray, almost diabolically, is his inability to shop for anything that doesn't involve horses, cattle, roping or tools for his shop.

A well-traveled worldly kind of cowboy I know has navigated remote ranches, big cities that even include San Antonio, South America, Europe, Canada and Japan. He cannot possibly find the toothpaste hidden in Wal-Mart.

This results in a pitiful situation where his bride does all the shopping even if he has just made a trip to town himself.

To further this travesty, he promotes his innocent lack of understanding about shopping by offering to help unload the groceries if she'll just wait an hour while he finishes his urgent task of, oh say, riding his horse.

Meanwhile, with milk, frozen food and perishables standing by, his bride knows he'll be right along as soon as it is all safely put away.

The same principal of innocence is offered if the cowboy has his eye set on a new horse that he is sure he needs to buy for his string.

Justification comes via generosity.

He will gift his bride with one of his current horses under the auspices of her needing an upgrade. He is more than willing to part with one of his prize steeds to help her out. That leaves him one horse short, and almost magically, a replacement will appear.

Sometimes this plotted horse trade will take months because his bride is not as thrilled with the idea as he seems to be. Often it means trading off her old dependable, trustworthy horse to some needy relative who simply cannot do without him. Again, this idea is his.

The trade often involves old women or children to add to the tender nature of the generosity. Tactfully, he will make his bride feel obligated to part with the security of her old horse for the betterment of mankind.

Cowboys and computers find a love-hate relationship where he cannot possibly pull up the bank statement for reconciliation but for a couple of years has been able to navigate ropinghorses.com with a knowledgeable dexterity.

The same guy that can mix complicated chemical formulas to spray brush and crops, and even fly the plane to put it on the land, will deny any ability to run a lawn mower, grocery cart, and certainly not the washing machine, dishwasher or microwave.

In the interest of full disclosure, the cowboy hero does offer some redeeming qualities. In the kitchen, he is completely willing to be in complete charge of Quality Control. Usually that entails sampling everything once, sometimes twice, most often in the case of pies.

Other valuable lessons for the cowboy's bride provided by the cowboy include:

- No matter how many exotic gourmet dishes you can make, cowboys prefer chicken fried steak, gravy, potatoes and beans to all the cuisines in the world.

- The best dessert in anybody's book is chocolate cake with gooey icing.

- You can always trust that the market will come up $20 to hit the break-even on any new set of cattle he wants to buy; trust him on that, he says.

- Always get on a fresh horse with his head in the corner so that he can't buck too hard.

- Never say to the wannabe, who might buy that unbroken colt, that his hat is on backwards.

- Always pick your spot with your back to the wind when holding herd.

There is an ongoing list of requirements for the proper care and feeding of your personal cowboy.

Recently, this worldly braniac cowboy claimed to not to know how to put mouse D-con in the barn. Some things just are not worth the fight.

CAUSE AND EFFECT

If you have not been in the cattle business, owned a cow dog or been married to a team roper, you may not believe this.

You know, however, that I am a basically truthful person and only arrange the facts when it seems appropriate.

Any head-of-household person who has managed to live a cattle-free life should be aware, though, that if ever anyone in the house would happen to take up team roping, it would encourage better housekeeping.

There is a connection to that train of thought and here it is as told to me.

"We finally got them all in the trailer and set out for home quite happy with our purchase. Once at our pens, we set all the gates with double-baling wire applied for safety and let these little darlings out into a very small catch pen.

"I have made the same mistake thousands of times and it never fails to result in disaster. I forgot to secure my cow dog. Ritadammit, the newer version of her name, takes her job seriously and decided to break the new steers into the routine around this outfit.

"Ritadammit becomes exclusively my dog at feeding time and any time she needs to be caught. Head cowboy's part is to yell at the top of his lungs, in very clear language, while I gather my pet.

"I got a little exercise in the pursuit but finally put Ritadammit in the cattle trailer. Now, this was a safe place for her. She was out of harm's way, couldn't get out and we could finish what we had started with the very wild and frightened new cattle.

"All was going quite well until I let Ritadammit out of the trailer. Completely insulted by her imprisonment, she made a beeline to the house.

"I thought that would be all right. She could just go sulk by herself while I washed out the trailer. These cattle were completely un-recovered from a current bout of scours. Took a while to clean the trailer but I got it done.

"When I went to the house, I saw the problem. It seems that Ritadammit had gone to her 'sulk' through the kitchen, through the dining room, through my office to see if there were any snacks on my desk, down the hall, into the bedroom and was lying in the middle of the bed.

"While in the trailer, she had become intimately involved with a good bit of the second-hand grass from the cattle. Actually, she had it on her feet, all over her coat and in her ears. This she had managed to track all through the house, on the tile and carpet and was doing a pretty good job on the bedspread.

"Consequently, the remainder of the day was given over to shoveling out the worst of it and mopping. I have to be honest and say it needed it anyway.

"This is just a word to the wise for people who have not been in the cattle business. Those of us who have will understand that this is no big deal.

"Just another day in paradise."

RODEOS, ROPERS & RIDERS

THE BUCKET LIST

Everybody, everywhere has owned a bucket, used a bucket or needed a bucket. Buckets through the ages have played a part in the very fiber of our lives, including "kicking the bucket."

We have the water bucket, milk bucket, mop bucket, slop bucket, coal bucket, ash bucket, grease bucket, feed bucket, lunch bucket, paint bucket and the ever popular, old oaken bucket. And a bucket of sorts, the chamber pot.

The ancestral poor-boy stories always include the lard bucket that became the lunch bucket. Lunch pail stories often include a long walk to school, uphill both ways and in the winter the sandwiches were frozen solid.

The bucket on the end of an old hemp rope strikes memories of a hand-dug well providing the only water on the place. The littlest kid, because he fit the best, got the summer job of climbing down the ladder to clean the silt from the bottom of the well.

My recall is that it was important that your horse was "bucket" broke so that you could carry a bucket of something on him while in the saddle. Naturally, the feed bucket was his favorite and he preferred it in front of him, full of oats. One horse was named "Smart Bucket" although I'm not really sure what that was about.

The milk bucket was the starting point of basic education for many youngsters. Every summer, based on the high nutritional value of milk, one lad was assigned to milk the cow to provide a never-ending supply of fresh milk while visiting his uncle's ranch.

The chore didn't have obvious "cowboy" value to the lessons he wanted to learn about riding, roping and punching cattle. Yet his uncle convinced him it was necessary to drink all this milk to maintain his strength and stamina to do the other work like fencing, pushing brush and breaking colts.

At the end of the summer, his last assignment was to turn the milk cow out to pasture. It seems the uncle didn't require as much good nutrition during the winter.

Buckets are good for many things that don't have anything to do with their intended use. Upside down to sit or stand on is the most popular. Kin to the bucket for functional "sittin' on" is the milk can.

At a recent down-home team roping that holds appeal to the "usta-be" cowboy set, a couple of portly cowboys were at the back of the chutes waiting their turn in the arena.

Good friends, the pair have found different challenges in reaching the stage of their life where the spirit is willing but the knees are stiff.

One continues to ride a 16-hand rope horse because he always did, but now getting on him represents a mission impossible. In remedy, he totes a milk can wherever he goes to use as a step.

His aging buddy finds the frequent need to sit down on something, anything, whatever is near. The milk can provides nicely for that as well.

The roping announcer called a name, putting one of them into action.

"Rusty, get up. I need my can."

"I just got set down here. I can't get up that quick," Rusty replied.

Call two from the announcer.

"Rusty! Get up. I've got to go rope, they'll turn my steer out."

"I'm trying," Rusty said. "Give me a hand here. My knees aren't working too good. This can is pretty comfortable."

During an understanding pause from the announcer, who had been clued in to the situation, the milk can was vacated and then used by its owner to climb aboard his horse.

As he backed in the box, they gave call three and Rusty ... well, Rusty had already settled back into his sittin' position on that handy milk can.

There comes a time in life when just everything looks like it needs sat on.

SOMETIMES THE BEST COWBOYS
ARE COWGIRLS

Every now and then, every cowboy, good or otherwise, will get a dose of reality delivered to him. The courier of this measure of humility is usually a cowgirl.

On the way to an all-girl ranch rodeo, the rig, fully loaded with horses and people, pulled in the yard just before dark. The resident ropers, champions in their own right, or mind, were at the arena spinning a few practice steers.

As all good hospitality demands, the gals were invited to come rope and the cowboys made every effort to be gentlemanly in their efforts to allow the damsels to have a turn at the roping cattle.

The first steer set the scene, scared the cowboys and reminded them "cowboy" comes in two genders.

These gals had been on the road all day, starting the trip and the beer drinking at 9 a.m., with a designated driver, also called "boyfriend." They rolled in, backed their horses out of the trailer, backed in the roping box and, more than handily, roped the fastest, dirtiest steer on the place before it got a third of the way down the arena.

They turned him, stretched him out, let him up, dusted off their hands and got another beer. Just a routine moment in another day in time with no extra effort involved.

The cowboys stood in stunned amazement, appropriate reverence and quite possibly, a little fear. A priceless moment for the cowgirl world.

Delicate and dainty aren't always descriptions given to such top hands, but beneath the chaps, dust and skill, lies the heart of a female who often struggles to maintain a thread of femininity.

For me, it was having nice nails – the kind you buy at the manicurist, which was the only kind I could claim for my own. My cowgirl friend and I were spending the summer trailing yearlings around the mesas of northern New Mexico at the direction of the boss, who also happened to be her husband.

We spent long days in the saddle, and did our best to "make a hand," but would also schedule a nail appointment every two weeks. It became a joke about "having the nicest nails in the cowboy crowd," not that you could ever see them under our gloves.

We knew we looked pretty rough and certainly were punchy enough to qualify our presence on most outfits. That was verified one morning just after daylight when we had gone to help a neighboring ranch ship a few semi-trucks of yearlings.

We unloaded our horses and headed to the backside of the shipping pasture along with a dozen other cowboys. When one puncher rode up next to me asked what outfit I was hired onto, I knew I had made the crossing from a get-in-the-way female to a qualified rag-tag cowboy.

The real marker of status in that summer came when a call for help came to provide a couple of cowboys to help a big outfit gather their cattle for shipping. Frank, the boss, took the call and was heard explaining he was already committed somewhere else that day, but he would send a couple of capable cowgirls to help.

What came next left the ranch foreman on the other end of the line laughing hard, and was likely a historical moment in time for the cowboy, cowgirl movement.

A pause in the conversation came when Frank said, "Wait a minute, first I'd better check and see if they have a nail appointment that day."

SALLY IS A GOOD OLE GIRL

Cowboying isn't just about riding a horse, swinging a rope or looking fine in Cinch jeans and a new George Strait straw hat.

Sally is a good ole girl. She's the kind you just like having around. Besides being beautiful, she is fun, rides well, and can sometimes catch both feet. She is also a favorite heeler for a number of headers in the team roping game because of those attributes.

She has an honest job and a little acreage of good grass for her practice cattle. Roping is her hobby and her cattle business has always been limited to however many cattle she needed for practice.

Last week, she was one head short when she did a cursory count of her holdings. She rode all the fences, rode the neighbors' country, called around and then finally rode her grassland again. All this handled on her high-dollar blue roan heeling horse.

She finally found the heifer hiding in an grove of oaks, head down and slobbering.

Any cowboy would have immediately diagnosed this condition as the ingestion of some weed that gave her a bellyache, put her in the pen, given her some hay and wished her well.

Sally hit the panic button. She called her favorite vet, a fun-spirited guy, who told her to give the heifer 30 cc of penicillin.

His phone diagnosis was that she probably had rabies and would be dead within the week and then they could send her head to the vet college. Additionally, everyone in the family and surrounding country would have to get rabies shots and likely be quarantined. Not thinking she would really believe him, he went on his innocent way.

Sally believed him. She went back to the pasture to get the heifer and bring her to the pens but couldn't get her to move. Finally she roped her, thinking this would be incentive to head the right direction. No luck.

She drug the heifer until the roan got tired. Then she tried again to tail her up and then had to drag her a little more. Finally, they got to the pens – the horse, the heifer and Sally were worn smooth out.

At this critical point in exhaustion, Sally's smart 12-year-old son shows up, looks over the situation and falls in with the vet. "Mama, I touched her. I have mad cow disease. I'm going to die."

After 30 cc of penicillin plus 15 cc of B12 just because, Sally had done all she could do for this poor, terminally ill heifer. By morning the medication had worked, the heifer was up, eating and walking the fence. Her neck was a little sore and she had a few grass stains on her sides, but otherwise, her life had been saved.

After such a successful doctoring event, Sally decided to worm all her practice cattle. Counting is also a critical cowboy skill, one that seems to have eluded her.

As she is telling her favorite team roping header about the event, she related that she had brought either 12 or 14 head in from a lease pasture, 7 or 8 from the roping catch pen and another couple from the trap at the house.

She wormed them, gave them all a B12 shot since that worked so well on the heifer and put them all out to winter grass.

As the tale is related, the cowboy listening does the mental math and comes up with 22-24 head of roping cattle, which Sally confirms to be "about right."

Wanting to check the numbers he says, "How many are left in the catch pen?"

She thinks, about 11.

"How many are in the house trap?"

She thinks, about 15.

"How many are in the lease pasture?"

She is sure she took back the ones that were there in the first place and this was either 14 or 16.

She is beautiful and she is a good heeler. However, just not everybody is cut out to handle the "cowboy" part of the job.

LIFE AFTER RODEO

Man wasn't created to spend his life locked in time and standing in one place forever. Time moves on and so does life and nothing drives that point home harder for the cowboy than the sport of rodeo.

In a flashback of the lifeline, he is a 20-year-old standing at the bucking chutes waiting for his bronc to be run into place. In a blink of an eye he is standing in the same place waiting for his 20-year-old son's bronc to do the same.

With a knowing eye you can look around any rodeo on any given day and spot the cowboys that "used to ride 'em." The sport calls them back to rodeo after rodeo, year after to year, to be part of something that once fed an inner calling at some point in their lives.

If they can't still be contestants, there remains a need to soak up the sights, sounds and smells created by those that are. From the broncs kicking in the chute during the national anthem opening to the last slamming gate signifying the end of the bull riding, it is all encapsulated into a familiar memory that plays like a recording through their entire life.

Many who once were heavily involved in the sport find other ways to stay tied to it. It is hard to find a stock contractor, announcer, rodeo judge, producer, rodeo secretary, stock hauler, tack salesman, etc. that was not at one time a contestant themselves.

When they no longer can compete in the sport, rodeo used-to-be's will sacrifice the adrenaline rush that comes with those precious few seconds in the arena to find new ways to let the sport touch them. Or they can become team ropers.

If they have kids they become the next generation on the road. Youth rodeos are continually raising and training a new segment of rodeo America.

If the kids are grown and gone or sometimes if they are not, "retired" competitors will find a job of one sort or another attached to the sport. For me it is behind a camera and with a keyboard. I photograph and write about it at every opportunity. It's not the same, but it is what I can do for the sport and for me.

As one former rodeo bull fighter recently told me, it's a hard trip down from the rush of the "in the arena" to standing behind a microphone announcing. I understood what he was saying but quipped to him that rodeo was finally paying me instead of me paying it.

Philosophical wisdom makes every attempt to sooth the beast within that was once the driving force behind the dedicated rodeo competitor. So today you get a little philosophy.

Life is like the coffee in our cups. It represents jobs, money, position, family and the other things we strive to collect to ourselves in a lifetime.

The cup, whether it is tin, ceramic, or of fine porcelain, simply contains the coffee but does not define the quality of it.

We have a favorite cup but usually we want a better cup. We frequently change the color, shape, size and personalize them. We have clever sayings and pictures printed on them.

We try to make a statement about ourselves by the cup from which we drink our coffee.

Being the humans that we are, we find places in our life where we concentrate only on the cup and fail to enjoy the coffee.

God help us all to not forget to enjoy the coffee each and every day of our lives.

CALL OF THE RODEO ROAD

Jess and Jim were legendary professional ropers, at least in their own minds and dreams.

In real life, they were ranch cowboys, thinking they'd work at that until the old rodeo road didn't look quite so long.

They practiced practically nonstop, as in, whenever the chance presented itself. A good bit of that time, this happened when they were supposed to be doing something else.

The boss had lined the crew out early that morning. The job was to sort off several loads of calves from the mother cows, getting them ready for shipping.

When the calves had all been cut, counted and safely stashed in the shipping pens, the two hotshot ropers looked over the cows.

Jess spotted what he was hunting for and hollered across the pen, "Look there, Jim, that old cow has a horn growing into the side of her head. We ought to do the boss a favor and trim that up before it gets to hurting her."

Of course, the good deed required roping the cow, but it was just part of their integrity to get the job done.

The remainder of the cowboys knew they would be the ones tailing over this big heavy cow, sawing off the bad horn, getting all bloody and very likely kicked in the process.

One of the hands came up with an old rusty saw from his toolbox. He made sure everybody understood it had been left there by somebody else since saws are not generally considered a cowboy accessory.

The cow was handily headed, heeled and the ropers were happy. The rest of the crew got her pulled over, the horn trimmed and the cow released.

While Jess was coiling up his rope, he spotted another one that might need a little trimming up, too. Before it was over, they had roped about half the cows. These all had to be tailed-down and something or other sawed off each one.

One of the cowboys who had been doing the cow wrestling told Jess he thought it was about time he started roping and Jess could saw off horns.

Jess gave him a shocked look and said, "Do I look like a carpenter?"

The fight was on, but when the dust settled, it had ended in a draw. It was also concluded that maybe the rest of the cows' horns would be all right until the next time they were worked.

A couple of days later, Jess and Jim were told to go check some yearling cattle down the road a ways. While they moseyed that way, they noticed a number of fancy yearling Hereford bulls in a nearby pasture.

An ample-sized pasture loaded with horned bovine. In the roping world, that is referred to as entrapment. No true roper could pass up that kind of challenge. Jess and Jim decided to take a little time out of their busy schedule and get some roping practice.

The bulls had weights at the tips of their horns to make them curve down properly as they grew. After the practice session, the ropers figured that these horn weights must have weakened the horns considerably. Several horns had broken, and some were just a little bent. They knew a bull never really needed fancy horns to get his job done and probably nobody would notice.

That Sunday, the paper had a big notice announcing a $5,000 reward for the culprits who had damaged the nearly priceless pedigreed bulls that were scheduled to go to the Fort Worth Stock Show.

Suddenly, that old rodeo road didn't look nearly so long. Jess and Jim make a quick decision that they would rather starve at team roping than spend time as guests of the county.

Practice doesn't always make perfect.

OLD GLORY WAVES ACROSS THE LAND TODAY

Old Glory will wave majestically in rodeo arenas across America this weekend, starting today. It's the Fourth of July and cowboys, if they are anything, are patriotic.

Don't misunderstand. It doesn't take a holiday for them to bring out the flag. It's there at every rodeo.

Honor to the stars and stripes happens first, before anything else.

Even the livestock seems to know the routine. Watch as the cowboys stand at the chutes, hats held over their hearts as the colors are posted and the national anthem is played.

The bucking horses in the chute will snort and kick the gate behind them like it should be part of the music's percussion.

For a rodeo contestant, it's an exciting sound that echoes in the recesses of their rodeo memories long after they no longer compete.

It goes with the smell of the arena dirt, the banging of the gates as livestock is moved around, arriving trailers rattling across the parking lot and the sound of hoofbeats as someone lopes a horse to the arena.

As all of us honor America, our freedoms, and the price paid for both, I find myself annually honoring the cowboy as well. This particular holiday is his "Cowboy Christmas," the most lucrative run of rodeos of the season.

Rodeo rigs are progressively bigger, fancier, and technology has kicked rodeoing up a notch from the days of standing in a pay phone booth to enter a rodeo or find out when you drew up. So much is different, yet so much is the same.

It still requires the basics. First, the cowboy has to get there, and second, he has to have brought his cowboy skills with him.

Fourth of July rodeoing is defined by road-weary cowboys, tired horses, pickups filled with dirty clothes, fast-food wrappers and muddy boots.

The dashboard of the vehicle is full of rumpled programs, Copenhagen cans, empty coffee cups, dust-covered sunglasses, gas receipts, a ball cap or two and a road map.

Many moons ago, when I was part of the rodeo world, I spent tired Fourth of July rodeo marathons wondering what the rest of the world did for their holiday and gave pause to the idea I might be missing something.

I'm not sure how any of that really works, I've just heard rumors about boating, fishing, barbequing and such.

Now that I no longer compete in rodeos, I still don't have the skills for the non-rodeo things. It wouldn't be true if I told you I didn't miss those days of driving across the state three times in four days to hit every rodeo possible.

I even miss the mucking around in the mud after a summer downpour at the rodeo grounds, washing off the barrel horse's legs and gear with the nearest water hose and cleaning up the kids, dog and my boots with the same effort.

I miss the camaraderie with the friends who shared the same passion for the same sport and questioned their sanity for it only briefly.

For me, it wouldn't be the Fourth of July if I wasn't standing in the hot sun, beating rain or dusty wind waiting for the next rodeo event to move the entertainment along.

So that's what I do. However, now I carry a camera and put what I know of rodeo in print.

I don't suppose I'll ever be anywhere else but at a rodeo grounds somewhere on the Fourth of July.

It just is who I am.

Join me at a rodeo for a look into the heart of the rodeo cowboy at his best. Today would be a good day to start.

BARREL RACERS – THE ARENA DARLINGS

A horse race of a different kind. That's barrel racing.

The event that was once open only to the ladies has found its own place in the major leagues beyond rodeo in co-ed futurities and other series events.

Through the ages, barrel racing, the clover-leaf pattern around three barrels, was heralded as the "beauty event" of rodeo.

While indeed the competitors represent the prettier side of the sport, they are no less committed to their event than those that pay big-buck entry fees to throw their rope in the dirt or have their head stuck in the same by a bull or bronc.

What appears to be hard-lined independence in these women is actually, more accurately, necessary capability. It takes a fair measure of intimidating grit to maintain the pace to keep self, horse and rodeo rig ready for the road.

While young boys were riding stick horses and wearing pot-metal pistols planning to be the Lone Ranger, girls in braces and braids dreamed of being a barrel racer. That included pretty clothes, fast horses and a cheering crowd as she raced through her pattern, riding hard to be the champion.

That dream grew to the reality that included a drained checking account, a four-legged sorrel standing in the corral eating a hole in her wallet with feed bills, vet bills and assorted expenses such as tack, a horseshoer and, of course, entry fees.

It has never been a secret to any barrel racer that barrel horses plot 24/7 to find a way to ruin your day. Winning the world in the practice pen on Friday can become dashed hopes the next day when that prize equine comes completely untrained at the rodeo.

When the champ runs by the first barrel like he doesn't see it, all there is left to do is begin with the rodeo queen wave as you loop the arena before making an exit out the gate.

Even though it is a highly frustrating sport because so much of the ability lies with the horse, barrel racing has done nothing but grow in numbers and popularity.

That happened in great proportions when a place for them to compete as a stand-alone sport was created. Barrel racers have never had the reputation of playing well with others.

Veterinarians and farriers are privy to the most demanding side of "can chasers." Vets will attest to the need for a degree in equine pediatrics as there isn't a horse in the world that gets babied more than a barrel horse.

Paranoia lives at the same address as every barrel racer. She can spot a limp, a cough, a twitch or a belly rumble in her horse before it even happens. And she lives in constant fear it will – just before she's supposed to be at a "big one" 500 miles away.

Ask any shoer who has put iron on a barrel horse how much "retraining" he received from the owner. When things go wrong in the arena, the shoer is at top of the list to get the blame, and the first phone call.

Every barrel racer carries with her on the road a complete veterinary pharmacy to ward off any possible ache, pain or ailment in her steed.

To find where she parked at the rodeo, one needs only to smell the air for a whiff of an assortment of liniments and secret concoctions surely cooked up in a cauldron.

On a more serious side, decades ago, Chris LeDoux recorded a song about barrel racers called, "Round and Round She Goes." In it he said, "Silver buckle dreams don't leave time for standing still."

The chorus summed up the spinning world of a down-the-road barrel racer.

> Round and round and round she goes
> Where she stops nobody knows.
> The miles are gettin' longer,
> The nights they never end.
> Old rodeos and livestock shows
> Keep the lady on the go.
> Lord, she loves to run those barrels,
> And it's the only life she knows.

There isn't a die-hard can chaser anywhere that doesn't identify with the truth in the song, or with the final lines:

> As she drove into the morning
> It slowly dawned on me
> How hard it is to tell a dream goodbye.

And so it is.

HOLIDAYS

COWBOY CATALOG SHOPPING

Topping this year's dynamic Christmas gifts, for the cowboy who has everything, is the latest innovation in roping practice equipment, the cooler dummy.

Combining two essentials for roping practice, there is now a plastic steer head that can be attached to any flat-top cooler eliminating the mess of a hay bale. We all know, you can't rope, either in dummy-roping practice or at the arena with the real cattle, without the ever-present cooler loaded with ice and a favorite adult beverage.

Of course, the catalog copywriters know just how to appeal to their market. "Use the cooler or tool box to store items in for travel and then pull it out to practice with or sit on at your destination."

This makes perfect sense if you are a roper. It also makes me laugh that the world of retail has found one more toy for the cowboy. There was a time when a cowboy was happy with a good whittling knife and worn-out rope.

Catalogs for the cowboy, especially the ropers, have blossomed from a few simple pages of saddle blankets, buckets and a stock offering of ropes to a glossy, burgeoning book that would rival the likes of a Sears and Roebuck catalog of old.

Cover to cover, you can order anything a cowboy might need from the sheets he sleeps on, decorated Western style of course, to the $30,000 trailer to haul his horse. Every attention to detail is given to the ropers' needs, including his reading material, his practice gear and his competition gear.

In case he needs chaps, chinks or a holster for his pistol, they are offered as well. His style of dress is catered to with selections of shirts, jeans, boots and even jewelry.

His dog, cat and horse can all be healthy with an assortment of veterinary items to select from and his stable can be clean and free of flies if he chooses to shop in that section of the book.

Perhaps a clearer picture that the direction the "cowboy" world is headed can be found in the catalog section that offers Western decorated cell phone holders, remote control holders and magazine racks, to hold roper supply catalogs and magazines, of course.

A complete section is dedicated to video, CDs and books on how to be a better horseman or roper. Every facet is covered in media selections ranging from teaching a colt his first lessons to the mental and physical preparation of the competition cowboy.

Not to be overlooked are the pages offering barrel racing, horsemanship and all kinds of roping clinics.

It is simple marketing genius to teach a few more recliner jockeys how to join the cowboy world and then provide them with absolutely everything they will ever need to look and be the cowboy of their dreams.

The world of "cowboy" supply is first creating the market, then catering to it completely.

And yes, there is even a "horses for sale" section. Let Fed Ex deliver your next roping horse to your door.

Christmas shopping for your favorite cowboy has never been easier.

GATHERING STRAYS TO SAM'S PLACE

Cowboys are all about strays. They round them up, rope, brand and doctor them, and in some mirrored reflection of the universe, you could say they become one and the same.

The dictionary defines a stray as a domestic animal wandering at large, homeless and without an owner. That pretty much sums up the cowboy with a question mark in the area of "domestic." Thanksgiving holiday in my world has become a gathering of strays.

The once solidly grounded-in-family-tradition celebration has migrated to a collection of eclectic folk all hoping to spend the day with friends doing something or nothing, whichever works.

Let's face it folks. The world has spun fast enough to scatter families to the wind and put hundreds and sometimes thousands of miles between their table and yours.

Easy travel, corporate employment and the lure of metropolitan paychecks have whisked away the kinfolk from their rural roots to suburbia. There they thrive with two and a half kids, a poofed and pedigreed dog, a boat and a feng shui backyard.

Buck Owens had a hit song in the late '60s with lyrics that said "There's always a party at Sam's Place, that's where the gang all hangs around." I'm headed to almost such a place this Thanksgiving.

While I don't expect to find Hootchy-kootchy Hattie from Cincinnati or Shimmy-Shakin' Tina who hails from Pasadena, I'm pretty sure Sally, the good ole girl from Stephenville, will be there to keep things beautiful and blonde.

Also in attendance at the turkey carving will be the crazy uncle, the class clown, the smart kids, a rodeo drifter or two and a couple of team roping partners who haven't yet found anyone else to rope with them or to invite them to dinner.

In the cowboy world of roping and rodeo, all gatherings begin with some sort of timed event, usually a roping. At a cowboy Thanksgiving dinner, it is expected that you'll bring along your horse and rope to finish out the day.

Dan, our favorite team roper hero, says that his family gatherings have always started this way. This works out well since his family is full of rodeo ropers of all ages, sizes and speeds.

However, this year the timed event was put on hold. Seems Granddad, who is in charge of the stock contracting, has, so far, only come up with one milk cow, a one-horned Hereford steer, a goat and two small donkeys.

Dan was mighty disappointed, as he has a brand new heel rope that he reports to be stiff enough to poke a cat out from under the trailer house. But with hope renewed, he'll head on down to "Sam's Place" and try this new nylon weapon out on a few unsuspecting Corrientes.

Thanksgiving will give many of us that opportunity in the true spirit of gratefulness for good friends and a bountiful table.

In the late afternoon sun, we will all waddle to the arena, moaning deliriously over the mental and physical memory of a magnificent meal.

If you can't be with the ones you love, love the ones you are with.

When I begin to recall the things for which I'm thankful, first on the list is life and the chance to experience joy and laughter.

Whether you spend your Thanksgiving with Mom, Pop and the cousins or quietly with the remote control, bag of Fritos and bean dip, my wish for you is that it is a joyful day.

Happy Thanksgiving from all the strays down at Sam's Place.

THANKFUL IS THE COWBOY WAY

A cowboy, as a general rule, is a man of few words. His thankfulness for his life is heartfelt but will be expressed in a simple manner. This is how it was explained to me.

The job doesn't pay much, but the air is clean. The benefit package is limited. Ranch rules are he can have two horses, one dog and he must use both for work. If he happens to get hurt or sick he will just have to get better and that would be sooner, rather than later.

His clothes don't have designer labels. He has one "town" shirt and, Lord willing, he will have enough money saved by Christmas for those new chaps he's been eyeing at the mercantile.

A pair of clean jeans, a mostly-ironed shirt and the dust knocked off the toe of his boots makes him ready for polite company.

He gets mail once in awhile. A latest catalog from the veterinary supply is a highlight in the week.

His schedule is pretty simple. It coincides with Mother Nature and Father Time. If the weather lets him and there is any daylight left, he will get it done.

His pickup is old but it still runs good. His horse is young and still bucks. For a cowboy, it doesn't get much better.

The roads out at the ranch don't have traffic lights and definitely no traffic jams. A traffic jam for a cowboy is when he is pushing a large herd of cattle through a small gate. The same gesturing might happen as with a jam of cars on an interstate in downtown Dallas.

Neighborhood gangs are made up of the folks from nearby ranches coming to help with the big work in the spring and fall.

The closest thing to smog arrives in the spring when a hot iron hits the dry hair of a baby calf. We call it branding smoke. Sometimes when he starts up the old pickup it belches a little black smoke. Some might call that smog.

Office politics don't exist and a nylon rope keeps things politically correct with a cow. It's been proven that there is a direct correlation between a cow's optic nerve and the tightness of a rope encouraging her vision to become more clear about where she's supposed to be going.

There are no lines to stand in to wait for anything. Back of the line to a cowboy means riding drag behind the herd. And they all have to do their time in that line.

A cowboy's outlook on the weather sums up in an ever-optimistic attitude. Every day is one day closer to the day it will rain. In the meantime, it's winter, and time to chop a little firewood before it gets dark.

In a single day, he sees more of creation than most will see in a lifetime of the Discovery Channel. He watches nature cycle in wildlife of all kinds as the coyote hunts, the deer and elk graze and the hawks on the wing observe from above.

For this life he is most thankful. He knows he can ride to the top of a ridge and be just about as close to God as he is going to get on this earth. His prayer for himself is that, Lord willing, he'll be here next year to say "thanks" once again.

THE CHRIST IN CHRISTMAS

At daylight on an icy, snowy Christmas morning, my dad went to the barn to do the usual daily chores. He was also keeping a secret there and the secret needed to be watered and fed.

Hidden in our barn was a coal black Shetland pony he had ended up with in one of his horse trades.

He had sold a perfectly good 2-year-old bay gelding for some Christmas cash and somehow ended up with this "prize" pony as part of the deal.

My dad hated ponies, believing that if you wanted to ride, you should ride a real horse and there were plenty of those around.

That point was driven home, literally, when the pony unloaded him on Christmas morning when he rode him bareback to the creek for water.

Landing hard on his jean pockets on the frozen ground left my dad with a broken tailbone that offered a painful reminder of his horse-trading abilities for months to follow.

While my dad provided many opportunities for memories during my formative years, there isn't a Christmas day I don't think about that incident and the many years that followed with the black pony adventures.

That simple, almost accidental, gift to us children became a memorable bookmark in our childhoods through many seasons.

I look at my children and wonder what parts of a tradition-filled holiday do they remember?

I'm sure there are individual stories for them, too, but generally, they remember the traditional things passed through generations of our family.

My teenage son tops his list with family get-togethers and big dinners. Food to fuel a growing boy's stomach is still a big part of his priority list.

However, with that is the delight in having the relatives gathered in one place.

My daughters recall the traditions they now carry on with their children. A cookie-decorating event, a family tree-trimming night, making grandma's recipe for homemade caramels and peanut brittle, the hanging of the stockings designed and sewed by grandma and the arranging of the traditional Christmas village.

A family favorite for generations has been the nativity display, complete with real straw to litter the barn floor and a light to represent the star in the East.

Bringing forth the solemn wonder of Christ's birth was, and is, as much part of our tradition as any one thing. Unlike the Christmas pony, it was not an accidental gift.

It is the one true gift that has kept on giving.

Political correctness makes every effort to sterilize the season by making it improper and, in places, even illegal to use the term "Merry Christmas." It is only a matter of time before they realize their "Happy Holidays" is only a version of "Happy Holy Days."

Somewhere in all the red and green everything, the masses of lights and never-ending glitter, it is important for us, as individuals, as a family and as a nation, to hold on to the true meaning of the season. The Christ in Christmas.

I never was very politically correct. Merry Christmas!

TIS THE SEASON FOR ROMANCE
FOR THE COWBOY'S WIFE

Romance is all about viewpoint. The level of romance is relative to the amount of time a rancher and his wife spend working together, often making little seem like more.

When you pick up the coffee table photo album, the photos are often of big brandings in a working pen with friends and neighbors hard at it.

That is not all there is to the action and the photo never reveals the endless hours of work that led to that particular Western Currier and Ives pictorial.

First, a date must be selected for a branding, usually a Saturday so the kids can be there to flank calves. All the neighbors and friends are called and the calendars marked.

The requisite vaccines, wormers and implants are ordered and the necessary equipment, i.e. branding irons, vaccine guns, sterilizing solutions, buckets, implant guns, knives for earmarking and castrating, and whetstones will all be checked to make sure they are in working order.

The cattle have to be gathered, sometimes done over a period of days, culminating in the final gathering to the pens on branding day.

The assignments will be well thought out who is going to rope, flank, give shots, brand and turn little bulls into little steers.

A long grocery list for beverages, snacks, coffee breaks, breakfast and lunch meals for a big crew will be written, shopped and stored.

All this is done ahead. By the time the assorted friend/neighbor crew shows up, the work is ready to commence.

A week before a branding can look like this:

Monday: Cattle are gathered from the big, rough, brushy pastures to smaller traps close to the working pens. The head cowboy, being good to his wife, will give her the choice of driving the feed truck with the clutch that slips or riding the green colt to push up the calves at the back of that herd.

At noon the same day, the wife mentions that she needs to go to town to get groceries for feeding the big crew. He will answer, "This afternoon we have to make another drive. We didn't gather all the cattle out of that pasture yet."

Making plans for the next day, the cowboy gives her instruction. "Tomorrow you better ride a broke horse, we're going to have to separate the bulls and move them to the back pastures to keep them from fighting and tearing down every fence on the place."

She recognizes the thoughtful caring in his consideration of the horse she should ride. She again mentions that she is going to need a day to go get groceries, since it's only 85 miles to town.

Tuesday: They saddle for the long trip to the back pasture where the bulls will go. He gives her a broke horse and decides he will ride a green colt. That will give the colt the needed miles of riding and guarantee that his wife will have to ride twice as hard doing whatever needs to be done. And, more importantly, with a good solid horse, she will be able to pack their lunches in the saddlebags.

Arriving back home late, it's too late to go to town for groceries, but she reminds him again she'll need time to do that and that it is 85 miles to town.

Wednesday: The cull cows are separated and driven to a separate trap for shipping later. Her good, broke horse is worn out from yesterday, so she gets her choice of the colts to hold herd on while he cuts the cows. Today the cowboy will give her lots of free, valuable instruction about how to get that colt to work a cow. When she seems less than appreciative of his wisdom, he lays it off to some hormone fluctuation.

At dark when the routine chores are done, she stands right in front of him and says, "I have to go to town no later than tomorrow to get groceries to feed this mob Saturday. You know it's 85 miles."

He replies, "We'll see. We have to go check waters tomorrow. Haven't done that for a couple of days, and we should probably go make sure the bulls are staying put."

Thursday: The bulls tore down a good stretch of fence to escape, the standpipe on one of the water reserves broke and the water valve that has been needing a little attention decided to finally go bad.
That evening his response to the grocery dilemma was: "Don't you have anything in the freezer you can just feed them?"

Friday: Bulls re-gathered, fences fixed, water valve patched and at noon, she gathers the checkbook, her list and heads out without saying anything to the other hat in the house. The last and only thing she had said to him since the previous evening was, "pass the butter."

He realizes she is going the 85 miles to town and to make good use of the effort he tells her, "While you're in town, pick up 250 doses of IBR and 250 doses of 7-Way, and 250 implants. We have 200 calves to work, but the women will probably give those things and you know they always waste some of the shots. While you are at it, you'd better get some 14-gauge needles, too. Those women always bend a lot of needles."

Silence.

Saturday: Depending on a gal's viewpoint and her proximity to the skillet, breakfast with 20 or so cowboys around the table can be fun. Funny stories, neighbors not seen through the winter months and plenty of good food start the day off best.

The anticipation of all the cowboy action when the sun comes up makes for a happy crowd. It is a strange but true phenomenon: Cowboys do not always regard cowboy activities as work.

When everyone has their fill of breakfast, the rancher will adjourn the crowd to begin the day's work. At this point, the ranch wife will often stay behind. She deals with the debris left from breakfast and proceeds to prepare coffee break snacks and drinks.

With that lined out, she will start lunch for the same crew that just left the table.

In the middle of the morning, one of the neighbors may notice the ranch wife is not around. "Hey, Jim, where's Karen today?"

Jim's sincere reply is, "I gave her the day off so she could cook." Truly, in his mind, that endearing gesture was all about love.

Mostly.

SADIE HAWKINS COWBOY-STYLE

Bachelorhood among cowboys is a confirmed asset to this free-livin' species of man.

I know one who is sure he's going to be a bachelor for his allotted three score and ten years.

He says, "Good judgment comes from experience, and experience comes from bad judgment, usually with bulls and women."

Not many things truly scare this cowboy. The cattle market strikes realistic caution and a genuine nervous moment comes with those dainty glasses on a stem – the kind that don't originally come with jelly in them.

A confirmed cowboy bachelor will avoid, at all costs, a woman with any kind of advantage. Those gals are bad enough straight up, but then someone declared a special day when they could legitimately propose marriage and, according to the rules, the male has to oblige. Sadie Hawkins Day is here.

For review, Sadie Hawkins Day was a daylong event in Al Capp's comic strip "Li'l Abner," named after Sadie Hawkins, "the homeliest gal in all them hills."

Each year on Sadie Hawkins Day, the unmarried women of Dogpatch pursued the single men.

If a woman caught a man and dragged him back to the starting line by sundown, he had to marry her.

Now it is celebrated only every four years, on Feb. 29.

The closest to Sadie Hawkins Day the cowboy world has come was with the invention of line dancing. It gave even the "homeliest gals" a dancing partner.

Our cowboy and his steady gal had been rocking along for a couple years going to local ropings, barbecues and calf fries. Occasionally, they would go dancing, and he would sometimes humor her along with a movie in town if she seemed a little antsy.

She was a good hand, knew cattle and good horses. He was quite happy with the current arrangement.

This cowboy had his life in order. The singlewide was paid for, there were roping horses in the pen, a friend he liked to rope with, a close Quik-Stop for menu selections and the girlfriend if he got lonesome or needed some errands run.

She, on the other hand, had another take on this deal.

Our cowgirl had been looking over the cowboy's life and could clearly see a few improvements that would help him out immeasurably.

He had been known to go for days eating only the gut-grenades available at the corner gas-and-go where you turn to go to the roping arena.

He had told her these were not unlike gyp water – a feller could get used to them if he just had enough intestinal fortitude.

He had also discovered that if you washed everything together, eventually your entire wardrobe matched with a light blue-jean hue, which was his favorite color.

With Sadie Hawkins Day approaching, she thought out a plan of action.

Her plot included dinner in town, her treat, at that classy joint that charges too much. Maybe a little wine and some soft Willie Nelson to set the mood. She would then explain the Sadie Hawkins Day rules and, gently, ask him to marry her.

They had been good friends, worked together without any serious injury, laughed a lot and were comfortable. She would spring this surprise and have him caught!

Our cowboy was not quite as slow as one might think.

Advised of the Sadie Hawkins Day mandate, he had been thinking about the situation and decided he had two options.

He didn't want to leave the country. So, true to the nature of a cowboy, he stepped up and took charge of the situation.

When they got to the high-class joint, he ate the biggest steak on the menu and for dessert he asked his girl to marry him – eventually, sometime.

With Sadie Hawkins Day safely handled, he let her pay the tab.

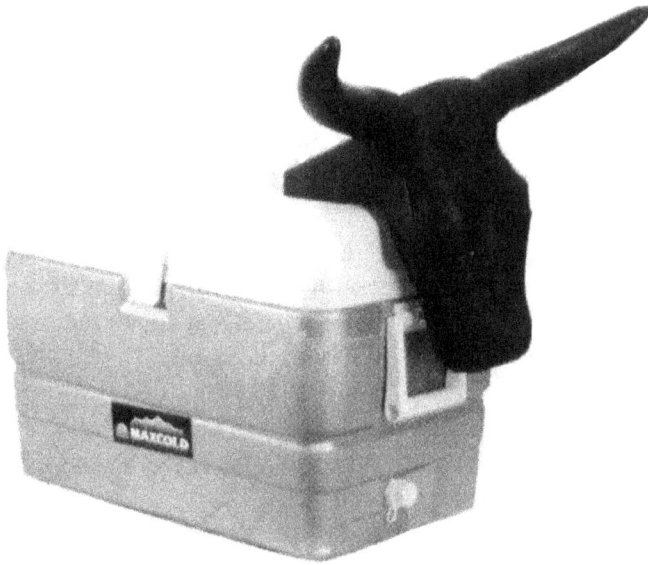

RURAL FUN

SUMMER SURVIVAL
AT THE ANNUAL FAMILY REUNION

Summers mark myriad events in people's lives, ranging from Disneyland vacations to weekend boating and fishing at the lake. Family reunions fit in there somewhere, and any of the sane ones that attend leave thinking, "What was I thinking?"

Family reunions are where a whole bunch of kinfolk, who never did like each other much, get together for a day or two and try to act like they are happy to be in the family.

Then they spend most of that time avoiding everybody they didn't like in the first place.

Every family has its "special people," and the last reunion I heard about began with the description, "Somebody needs to get the net after that entire bunch."

Many reunions are fodder for a sitcom script. Here is just one.

The clan runs the gamut of all kinds of crazy.

The cousin that is the judge mistook the rest of the kinfolk for voters and told stories and jokes nonstop.

His motto was, "Any story worth telling is worth adding a little something to."

The family dictator – a.k.a. the one who organized the event in a resort town one state over from where they all lived – instructed one cousin he was in charge of Sunday morning breakfast.

Her list was for him to get 80 eggs, 5 pounds of sausage and 5 pounds of bacon.

When two of the expected clan didn't arrive, she cut the number to four dozen eggs with the same sausage and bacon.

No one was quite sure who it was that didn't show, but they knew they must be egg-eating dudes.

The fact that the organizer was a schoolteacher made this math somewhat concerning but then someone recalled she was allowed to teach only special-ed students.

Uncle Mike was a big winner at the horse races and told the clan he would treat them all to a drink at the casino.

"You know, when you hit it big at the races, a feller can do a lot of things," he said.

"How much did you win, Mike?"

"Twenty-three dollars," he answered.

The old uncle, who is 84, has a young steady girlfriend of Latin descent, and he spent the weekend giving tango lessons on a spontaneous basis.

A whispered warning passed from cousin to cousin to be sure and not ask the old guy about his love life unless you were prepared to hear more than you wanted about sex at 84.

Touristy events included a trip to a local Western museum that made an impact at some level on most of them.

A good number of them opted for a trip to a Friday night buffet, highlighted by one cowboy landing, passed-out, on a stack of clean glasses in the pantry while a security guard babysat him until his wife was located.

Although classified pretty much as hillbillies through and through, there were a few that made every attempt at being civilized, even just for the weekend.

One woman proudly spent $62 on a pedicure, foot massage, had little daisies painted on her toes and her skin twinkled from the sparkles in the lotion that was applied.

Another lost 11 cents playing the penny slots and whined about it for two days.

The jewelry-peddler cousin worked throughout the event and sold a load of stuff to the kinfolk and even to a few local shops.

She hauled all the females down to a trendy boutique where they bought a $150 shirt or two.

Others thought the $5 price to see the museum was good enough and, besides, they already had a shirt.

RETURN OF THE HILLBILLY FAMILY REUNION

Anyone who has ever had a family has had to endure the occasional family reunion.

Last summer, I bravely ventured out to somebody else's family reunion. I wrote about this family when they gathered the previous year, but I had never personally met any of them, except the one that invited me.

I met the 85-year-old uncle, who last year was in the company of a young Latin tango dancer.

This year he was traveling solo, but had lost none of his will to tell tales of his sex-at-85 adventures should one be unwise enough to inquire about his love life. Most of the time, he could be found hovering over the platter of bacon-wrapped shrimp pondering their possible aphrodisiac properties.

The cousin who is a district judge counts family among his friends and voters. Campaign donations are always the topic when deciding if favors of any kind are to be granted, including refreshing your glass of ice tea.

Last year's event to this mountain resort landed them in some rustic cabins where they charged one cousin with providing 80 eggs and 10 pounds of breakfast meat. A cook-your-own sort of place.

This year, the group landed in a high-class hotel with what I heard referenced as "whorehouse prices," but it included a complimentary hot breakfast.

The judge and his family, along with his brother and family, were housed in a top-floor executive suite.

These men are rough-country ranchers from the Palo Duro area, raised up poor and hard-working.

The one brother stayed on to run the ranch after college, while the other became a progressive embarrassment to the family by first becoming a lawyer, then a politician on his way to the judgeship.

When the rains came, the roof of this posh hotel began to leak in the executive suite.

This would have been a tremendous problem for most who were paying through the nose for the privilege, but not these hillbillies. They just moved the beer cooler over to the spot, flipped the lid open and let it catch the rain.

Hoping to class up a few of the kinfolk, the planners of the soirée booked Asian cooking lessons from a local culinary chef.

Any previous knowledge of this particular ethnic food included the little-known fact that the Chinese restaurant in Hereford, Texas, next door to the laundry, utilized the discarded starch water to thicken their gravy.

The story goes that even the calf-fry-off-the-branding-pot eaters needed several Mimosas to survive the chicken deboning demonstration.

One gal that married into this bunch finds great relief in their "normalcy," as compared to her family.

She recalls some of her own kinfolk. Just for starters, there is the one with the long hair and tattoos who makes a living playing acid rock, or when that isn't good, is an operator for underwater seismographic oil exploration.

Another, with the same hairstyle and a law degree, makes his money picking a banjo. Yet another raises doves and, when the eggs don't hatch, works at Wal-Mart to make ends meet.

Thankfully, she has seen nary a one of them in years, except the cousin that showed up with a couple of horses and a pig when everyone was making a mass exodus from the coast to escape Hurricane Katrina.

Writing about other people's kinfolk sometimes makes me miss my own, but not enough to wish for a family reunion.

PIGS, KIDS AND HOW TO MEET GIRLS

The cell phone rang and, when answered, the phrase "I'm on my way to a pig sale," could be heard over the roar of the F-350 flatbed pickup as it rolled down the highway.

The digital hi-fi wireless age has allowed people a glimpse into our personal lives and let them find us in places it never occurred to them we would be.

This particular mother was headed out with her husband and son to get new livestock for this year's county fair projects. The guy calling was her computer networking tech.

He was someone from her other life. The one where she is a business owner in skirts demanding efficiency, order and perfection.

None of which she would find at the pig sale.

He politely tried to reason why in the world would she be going to a pig sale?

It is sometimes a little hard for "regular" people to grasp the concept of livestock projects for FFA and 4-H youth headed to the county and state fair.

Their deductive reasoning finds no logic in driving long distances to buy just the right pig or three, spending the next three months buying feed and dispensing it to said animals knowing full well the highest odds are for losing money in the end.

The value of this effort year after year is not tangible because it really isn't about the pig. It isn't about how much he ate, how well he showed or how many cents per pound he'll bring in August. It is all about the kid.

It is about an investment in responsibility. From now until fair time, the kid will be required to daily feed, groom, exercise and clean-up after a pig. (Or other show animals, but for the purpose of the lesson, we'll stick with the pig.)

And the most surprising part to those same amazed regular people is that the kid really does like it and so does the pig.

Something happens within the youngster when that animal becomes part of his daily life and dependent upon him.

Psycho-babblers would call it bonding. The pig has no idea about that but he knows that the short person that shows up every day is bringing chow and some attention. That's good enough for a pig.

Just about the time the pig and the kid are into a steady routine and summer seems endless, it is fair time. Again, although it's the pig that gets the kid to the fair, it's still really not about the pig.

Pig pens are cleaned, assigned and move-in day is a flurry of squeals and snorts and some of those from the pigs.

Kids are running, jumping, climbing, laughing and joking. You'd think they'd just arrived in Disneyland Instead of the pig barn at the fair grounds.

The teen kids stand in segregated groups with mirrored sunglasses, cool ball caps, bling-bling jewelry and perfect lip-gloss trying to pretend they don't notice each other. Many a long-term marriage's courtship began in such a place – a week in the county fair pig barn.

All the kids, large and small, are giddy with anticipation of the fun they'll have for this one week. Any doubts they may have had in April about whether they wanted to show livestock this year have vaporized into sights and sounds of the pig barn.

My 13-year-old son had some of those momentary doubts. But they quickly vanished when his reasoning skills kicked in.

I was assured of that when he said, "Mom, the fair is a good place to meet girls."

THE COUNTY FAIR BELLES OF THE BARN

Every day, sometimes twice-a-day, the livestock barns at the county fair see a transformation that seems almost magical.

The ladies of the livestock pens transform from normal teens in a pig barn to bling-bling fashion queens headed to the show ring.

The full-house crowds that always gather around the show ring prior to the junior livestock shows at the fair are missing out on the real show that takes place in the barns in preparation for the competition.

While there are at least an equal number of male exhibitors, the transformation for them is less dramatic and more low-maintenance, so for the purpose of this story, I'm telling the "girl" version.

A family testing period takes place for hours prior to "showtime." One or both parents of an exhibitor is up to their whatevers in suds and water getting a pig, who is not known for his hygiene, fluffed, buffed and show-sheened for competition.

All this happens while the sole proprietor of the pig is putting on makeup, spraying down loose strands of hair, giving a permanent set to ringlet ponytail curls with a color-coordinated ribbon and polishing up the rhinestones on her hoop earrings and belt.

A short time before this, she was simply a normal teen in a T-shirt and jeans.

While the bling-bling princesses are parading the show ring looking like runway models from a rural setting, their moms try to watch from ringside.

They aren't hard to sort out from the crowd. They have a towel slung over one shoulder, are wearing high-top rubber boots, old jeans and a shirt splattered with water and other stuff not discussable, and they carry at least one, if not two, water spray bottles.

Tendrils of their hair fall around their flushed, makeup-less faces. Anxiousness keeps them on the move with a desire to watch the competition while knowing they have yet another hog to wash before the next class of competition.

Most of these moms were once show ring darlings themselves. Those few short years of moments in the spotlight at the county fair are a distant but poignant memory.

It is now what keeps them doing what they do – passing the torch to their sons and daughters knowing the experience is more than just fun, but a solid foundation for life-long attitudes and character traits.

While Dad is often the muscle, the coach and stoic competitive instructor, Mom is the organizer, the cheerleader, the motivator and of course, in charge of all lists for all things before, during and after the fair.

Somehow in the midst of the exhaustion from the preparation to get to the fair with all the necessities for providing for a family and critters for a week away from home, they maintain the ability to smile, and smile, and smile some more.

After all, it's County Fair week and by Saturday night there will be many who know they had a lot of fun – but they will have to rest up a couple weeks before they can recall just when it was that they had fun.

And Belles of the Barn?

They start young, but the true queens are the moms who, year after year, give of themselves to raise another generation of kids who are not just about the beauty and the bling.

They know that what really counts is the beauty that comes from hard work, self-sacrifice and responsibility to other people and to their livestock.

Come to the fair and I promise you will see what I mean.

EPIC CLOTHESLINE NOSTALGIA

On a recent walk through my yard on the way to peer over the back fence, I ducked under the clothesline.

There are a couple of generations of us left that do that without a given thought. It's there. We duck.

However, this time, something occurred to me about that old clothesline.

I've raised children who now have children and none of them have used a clothesline.

If they lived anywhere there was one, it might have served to throw a rug over or tie the dog to, but never did they experience life's connection offered by a clothesline.

Never did they know about wiping off the lines with a cloth before hanging up the freshly washed clothes.

They didn't learn clothesline etiquette about hanging whites with whites and preferably the undies went on the line behind the bed sheets for propriety's sake.

And those bath towels, like drying off with sandpaper after a day in the sun and wind! I have to admit though, crawling into bed with fresh sheets that were sundried is aromatic heaven.

Shirts were hung by the tail, never by the shoulders. And the neighbors, if you had any, might raise an eyebrow if the whites were a dingy gray. Remember Mrs. Stewart's Bluing? I never did understand why putting blue stuff in the water would make the white clothes whiter.

Clothespins hung in a bag that stayed at the clothesline. It was a big upgrade when the pins went from the wooden peg-type to the ones with metal springs in them. That unleashed a whole new industry among inventive children who used them as fire-throwing catapults, mousetraps, rubber band shooters and detonators.

One could always tell how big the family was and the basic ages of the children by what hung on the clothesline.

The graduated sizes of jeans were always a clue, hung next to the shirts that went from Dad's denim work shirts to little striped T-shirts that looked like they'd been worn by several siblings through the years.

Socks were matched up before hanging by the toes so they could easily be folded or rolled when taken off the line.

And jean stretchers? Remember wrestling with those metal frames stuffed into Levi pant legs to form a stout crease and minimize ironing? And make sure you got that crease dead-on straight.

I still have visions of my mother racing the rain that was blowing down off the mountain, pushing dirt clouds ahead of it and promising to ruin the efforts of her day-long washing project.

Grabbing huge armloads of clothes in one fell swoop, clothespins flying, she'd move to rescue, first, the sheets and white things that would most show the effect of the muddied drops.

And then there was winter. Our ranch house was at 8,200 feet in elevation and summer was very short. Therefore, for more months than not, clothes freezing on the line was the bigger problem.

In winter, it wasn't uncommon to have clothes, arriving frozen stiff from the line, strung around the dining room, hung on chairs and anything else that didn't move. And there was that flimsy folding rack that sat near the only source of heat in the main part of the house. I recall it full of drying diapers, long before the day of disposables.

Most of America has seen the last of these things. There remain a few that still hang on a line when they can.

Propped up in the middle by a two-by-four post, the old sagging lines still offer a friendly greeting to a home where clothes are cared for by love, not by Whirlpool.

'TIS THE SEASON FOR CAMO AND AMMO

If big-game hunting and all that goes with it offends you, I will understand if you want to turn the page. It's not for everybody.

On the other hand, there is a world full of bodies that thrive on the hunt with every intent for the kill and none of their reasons are the same.

The fall months bring out the masses of hunters seeking trophy antelope, deer and elk, and the world turns camouflage.

Drivers dressed in camo, the highways are covered up with trucks loaded down with 4-wheelers, coolers and camping gear. Some of them even bring a gun.

At a grocery store in a small town on the perimeter of hunting land, a guy leaving the store made a comment that he had been going to go in and buy a loaf of bread. He indicated there were so many hunters inside he figured it would be quicker to just go home and make biscuits.

Masses of businesses depend on the hunting industry for a major boost to their economy. Small communities in hunting country have signs everywhere offering free chili suppers and free hunter breakfasts.

One guy reported it cost him about $200 in gas to go around to all the little towns and eat their food. He isn't a hunter, just an eater.

Posted signs read 45 miles to the next ammo store and if they don t say "ammo," they say "beer."

I come from a long line of meat hunters who, indeed, hunted, but foremost for family sustenance and somewhat of a right of passage to manhood for the growing boys in the bunch.

The locals simply gear up, go kill something, bring it home, skin it out, cut it up and know they have winter meat in the freezer.

While they have plenty of fun doing it, as witnessed by the masses gathered around the skinning tree, it is more a way of life than an epiphany.

When the fall weather turns cool, the primal instinct to hunt and kill a little winter meat rises like sap in a maple tree. It is one of the few things that bring out the same stalk, kill and drag-it-home instinct in men the same as it did in the days they lived in caves.

Die-hard hunters look offended if you ask them, "Are you going hunting this year?" In their minds, it is a national holiday. Hunting season is marked on the calendar first and before anything else for the year. Of course, they are going hunting. Any more silly questions?

I've known men to quit their job so they could go hunting. Another who would injure himself just enough to qualify for some paid time off which, of course, he used to go hunting. Neanderthal instincts are not buried very deep in most men.

My favorite story from this year's hunt was when my daughter telephoned via cell phone from the hunting grounds to announce success. She then emailed me a picture, again via cell phone, of the prize animal with the hunter and his bow.

Nothing is safe from technology.

ABOUT THE AUTHOR

Julie Carter paints cowboy pictures with wit and words in her attempt to bring a unique point of view to cowboy life - be it cowgirl woes, cowboy behavior, rodeo adventures or simple essays about the nature of the enigma called "cowboy."

From her ranch raising in the mountains of southern Colorado to her ranch living in New Mexico and a few hundred rodeo arenas in between, she writes with the intent to make you laugh, make you think or a little of both.

Her stories are those she has experienced herself or as told to her by others. The lifestyle and experiences are not unique to her, but Julie writes as the voice of those that came before her and those that still live down dirt roads far behind the cattle guard.

Julie writes a weekly syndicated column, *Cowgirl Sass and Savvy*. She is also an award-winning photographer. Photographs within this book are by the author.

Visit Julie's website at www.julie-carter.com or contact her at julie@julie-carter.com.

www.ingramcontent.com/pod-product-compliance
Lightning Source LLC
Chambersburg PA
CBHW051825090426
42736CB00011B/1657